Michael Kitchens

ESTIMATING and PROJECT MANAGEMENT for BUILDING CONTRACTORS

Published by
ASCE Press
American Society of Civil Engineers
345 East 47th Street
New York, New York 10017-2398

Abstract:

This book provides guidelines to the day-to-day processes of construction estimating and project management. It is a useful overview for both beginners and seasoned professionals who seek a better understanding of the skills and abilities required of a good estimator and project manager. These include decision-making skills; knowledge of labor, equipment, contracting and subcontracting, financial considerations and cost control, job site safety, schedules, claims, delays, quality assurance; and more. This book also provides information on ethics, training, computers, and real-life experiences.

Library of Congress Cataloging-in-Publication Data

Kitchens, Michael.
    Estimating and project management for building contractors / by Michael Kitchens.
        p.   cm.
    Includes bibliographical references and index.
    ISBN 0-7844-0148-9
    1. Building—Estimates—United States. 2. Building—United States—Superintendence. I. Title.
TH435.K495 1996   95-48347
692'.5—dc20   CIP

      The material presented in this publication has been prepared in accordance with generally recognized engineering principles and practices, and is for general information only. This information should not be used without first securing competent advice with respect to its suitability for any general or specific application.
      The contents of this publication are not intended to be and should not be construed to be a standard of the American Society of Civil Engineers (ASCE) and are not intended for use as a reference in purchase specifications, contracts, regulations, statutes, or any other legal document.
      No reference made in this publication to any specific method, product, process, or service constitutes or implies an endorsement, recommendation, or warranty thereof by ASCE.
      ASCE makes no representation or warranty of any kind, whether express or implied, concerning the accuracy, completeness, suitability, or utility of any information, apparatus, product, or process discussed in this publication, and assumes no liability therefore.
      Anyone utilizing this information assumes all liability arising
from such use, including but not limited to infringement of any patent or patents.

Photocopies. Authorization to photocopy material for internal or personal use under circumstances not falling within the fair use provisions of the Copyright Act is granted by ASCE to libraries and other users registered with the Copyright Clearance Center (CCC) Transactional Reporting Service, provided that the base fee of $4.00 per article plus $.50 per page is paid directly to CCC, 222 Rosewood Drive, Danvers, MA 01923. The identification for ASCE Books is 0-7844-0148-9/96 $4.00 + $.50 per page. Requests for special permission or bulk copying should be addressed to Permissions & Copyright Dept., ASCE.

Copyright © 1996 by the American Society of Civil Engineers,
 All Rights Reserved.
Library of Congress Catalog Card No.: 95-48347
ISBN 0-7844-0148-9
Manufactured in the United States of America.

Cover photograph courtesy of the Bay Area Rapid Transit District.

Dedication

This book is dedicated first and foremost to my family - Judy, my wife, and Crystal, Kylee, and Cody, my children.

This book is also dedicated to my mentors in the fields of estimating, project management and management philosophy. Those mentors are:

        Rudy Alvarado

        Dorman Blaine

        Gene Carlier, P.E.

        Bill Judd

TABLE OF CONTENTS

	Introduction	1
Chapter 1	**The State of the Construction Industry**	3
	Business Statistics	3
	The Risks	4
	The Appeal (The Rewards)	6
Chapter 2	**Estimating and Project Management Roles**	7
	The Estimator	7
	The Project Manager	9
	The Project Superintendent	11
Chapter 3	**The Decision to Bid**	13
	Contracting Opportunities	13
	Availability of Bid Documents	14
	Participation of Surety Company	15
Chapter 4	**Preparing for the Estimate**	20
	Ensuring Subcontractor & Supplier Coverage	20
	Bid Document Review	21
	Site Investigation	26
	Pre-Bid Conference	30
Chapter 5	**Preparing the Estimate**	31
	The Quantity Survey	31
	Estimator's Database	34
	Labor Rates	37
	Pricing the Labor	43
	Material Costs	48
	Equipment Pricing	49
	Unit Prices	53
	Initial Schedule	55
	General Conditions Costs	59
	Insurance Considerations	65
	Testing Costs	69

Chapter 6	**Bid Day Procedures**	70
	Preparing for Bid Day	70
	Addenda	71
	Bid Bucket	72
	The Estimate Spreadsheet	73
	Receiving and Evaluating Subcontractor and Supplier Bids	77
	Scope of Work Checklists	80
	Profit Considerations	87
	Taxes	90
	Costs That Should Be Considered	90
Chapter 7	**After the Bid**	95
	Post Mortem	95
	The Low Bidder	95
	The Project Management Team	96
	The Coordination Meeting	97
Chapter 8	**The Paperwork**	100
	Project Files	100
	Scheduling	101
	The Buy Out	104
	Subcontractors	107
	Purchase Orders	112
	Submittals	115
	Requests for Information/Clarification	119
	Job Site Safety	121
	Job Site Housekeeping	124
	Cost Control	124
	Project Meetings	145
	Project Documentation	146
	Progress Billings and Payments	150
	Quality Assurance and Control	158
	Project Close-Out	159

| Chapter 9 | **Contractual Considerations** | 163 |

The Contract 163
Change Orders 168
Claims 171
Delays 174
Notices 177
Mechanic's Liens 177

Chapter 10 Other Topics 188

Errors in the Estimating Process 188
Common Errors in Project Management 190
Computers in Construction 191
Ethics in Construction 193
Training 196

Chapter 11 Managing Project Risks 200

Chapter 12 Lessons to be Learned 207

Case Studies 207

Glossary 212
Bibliography 232
Index 237

INTRODUCTION

The construction industry is one of the largest industries in the United States of America. Construction accounts for approximately 10% of the nation's gross national product and about 15% of the country's employment. It is the largest user of steel, aluminum, copper, cement, rubber, lumber, brick, building supplies, fuel, power, and a multitude of other products that rely on the construction industry.

The construction industry has almost 1,000,000 participants; however, more than 60% of the contractors have three or fewer full-time employees. In the finest tradition of capitalism and free enterprise, construction contractors are fiercely competitive with one another.

The construction industry has become increasingly complex through the years as a result of technological advance, natural evolution and litigation. However, unlike other service industries (the construction industry is unquestionably a service industry), the success of a construction business enterprise is much more reliant on the caliber and attributes of its people than it is on technological advances or by the availability of capital facilities. The construction industry is certainly, above all else, a people profession.

The economic turnaround of the United States during the last few years has been devastating to the construction industry. As a result of reduced demand for construction services, an already competitive industry has become even more competitive which directly relates to declining profit margins and sometimes bankruptcy.

In 1978 twenty-two out of every ten thousand general contracting businesses failed. By 1992, that number had increased over six-fold to 140. This increase is due to several reasons.

- In the heat of competition, contractors are bidding at or even below their costs.
- Contractors are not providing proper training to key personnel.
- Contractors are taking more risks with less reward than ever before.
- Contractors as a whole are not staying current with the latest technology.
- Contractors do not stay abreast of new management philosophies.

Other industries are feeling the brunt of the recession as well as pressure from other nation's producers in what is quickly generating into a world economy. These other industries are tackling the onslaught through improved management techniques, such as total quality management, plan-do-study-act, reengineering, performance excellence and quality excellence, to name a few of the quality enhancing programs.

Contractors must become more sophisticated and knowledgeable business managers. The old-fashioned management style of "just do it and we'll count the money later" must give way to contractors who are committed to the management techniques that other industries are embracing.

The initial step involves reviewing the processes currently in use in the industry. One of the primary purposes of this book is to provide the construction industry with both a mirror to reflect current thinking and a crystal ball to show where improvements are readily achievable in the processes of estimating and construction project management.

In the construction industry, change is too often viewed as a one-time event. That view is very short-sighted. Change is not an event; it is an ongoing process. This is especially true in the construction industry. Better products, more efficient equipment, different materials, etc., are being developed daily. However, the processes of estimating and project management, the foundations of the construction industry, are changing at a snail's pace compared to the foundations of other industries.

Other industries are embracing new management techniques which result in empowering and training employees toward the goal of continual improvement.

The research and experience of the author point to the fact that in order for the construction industry to begin to step into the age of continuous improvement, cultural and behavioral changes are required. Some of those changes include:
- Encouraging open and honest communications between companies.
- Encouraging sharing of bad experiences.
- Having less of an independent attitude.
- Providing employees with more and better training opportunities.
- Having less of an adversarial/competitive attitude.
- Focusing more on bottom-line results.
- Improving focus on the risks inherent in the construction industry.

The purpose of this book is to provide a first step towards incremental improvement.

Chapter 1

THE STATE OF THE CONSTRUCTION INDUSTRY

BUSINESS STATISTICS

As previously stated, construction is big business in the United States. The U.S. Department of Commerce - Bureau of the Census publishes statistics every five years which provide data relative to various business types. The latest data available is for the year 1987.

In 1987, there were 157,600 general building contractors with 1,278,000 employees generating a payroll of $27.8 billion and producing revenue of $212.63 billion. The average general contractor in 1987 had 8.1 employees and generated $1,349,169 in sales.

The values for 1977, 1982 and 1987 are listed in Table 1.1 below.

Table 1.1 General Contractor Averages		
Year	No. of Employees	Annual Revenue
1977	4.12	$342,682
1982	9.00	$688,077
1987	8.12	$1,349,169
1992	-	$1,317,464

According to the 1995 Almanac of Business and Industrial Financial Ratios, in 1992 160,232 general contractors generated $211.1 billion in revenues, an average of $1,317,464 annually. From 1987 this is a reduction of growth.

In 1966 the construction industry produced 11.9% of the GNP (Gross National Product), by 1987 that number was 8.9%, and by 1994 it had further reduced to 7.9%.

In 1978, 22 out of every ten thousand general contractors failed. By 1992 that number had risen over six-fold to 140 out of every ten thousand.

In 1970 only 1,687 construction-related businesses failed. General contractor failures alone exceeded 4,500 in 1992.

In the construction industry only 5% of the businesses generate revenues exceeding $2,500,000 annually. This indicates that many of the businesses are solely owned or have working partners.

For 1992 the average before-tax profit for a general contractor was 1.2%. The average return on equity was 8%. (These averages do not include the failed contractors). Thus for 1992 the average contractor earned $16,881 before taxes. The typical profit included in the bid for a $500,000 project ranges from 2% to 5% for the general contractor.

THE RISKS

As evidenced by the 636% increase in the rate of general contractor failures over the past fifteen years, the risks involved with the construction industry have grown exponentially. In the context of the construction industry risk may be defined as a financial exposure that is not factored into the cost of doing business. Risks associated with the construction industry are listed below.

- The very nature of the competitive bid process forces mistakes which range from minor to catastrophic.
- During the past two decades owners and their attorneys have become experts at including exculpatory (risk-shifting) clauses in owner/contractor agreements.

  Examples of exculpatory clauses include the following:
  - No damages for delay clause.
  - Site investigation clause requiring the contractor to visit the site and ascertain the character of existing conditions for themselves.
  - Liquidated damages clauses.
  - Indemnity and hold harmless clauses.
- Project financing lenders tend to participate more aggressively in construction projects.
- With the current state of the economy and financial condition of owners, failure to fund the contractor's completed work has become a major problem in the industry.
- During the construction phase the probability that costs will exceed budgeted amounts is high. (The before-tax profit is much less than the bid profit as a percentage of volume)

- General contractors typically rely on subcontractors and suppliers to furnish and/or install 70% to 80% of the value of a construction project. Thus, in reality the subcontracting industry has true control of the success or failure of a project. Lack of funds on the part of subcontractors place general contractors in the position of having to complete subcontractors' work, usually at a much higher cost when subcontractors fail.

- The construction industry is high risk with regards to injuries. Although worker's compensation insurance mitigates a contractor's damages, many injured workers file civil lawsuits against the general contractor and a host of other project-related personnel.

- Project schedules and milestones are developed by the owner but the contractor adopts the liability of schedule performance when he accepts the job.

- Most construction contracts shift the risk for subsurface-related conditions and extra costs to the contractor.

- The contractor must contend with environmental laws that are surfacing or changing regularly.

- The contractor is at the mercy of local building officials to perform timely inspections.

- All contractors require surety credit to sustain their businesses; however, bonding companies are becoming more conservative in providing bonds to contractors.

A look into the crystal ball for construction reveals a potential for risk which has either not yet been reached or has occurred in very limited areas. Some of those potential risks include the following.

- The subcontracting industry is making headway to enact legislation in various states that would require the owner to set up escrow accounts from which individual payments would be administered for each entity providing materials and/or labor to the project. One of the hammers a general contractor has over subcontractors to guarantee performance, reliability and quality is that the general contractor has the subcontractor's money. This advantage would be neutralized with the aforementioned type of legislation.

- Owners and their attorneys are getting wiser with respect to mechanic's lien laws. As such, there is a movement to refuse to accept conditional lien releases and require the general contractor to pay suppliers and subcontractors prior to receiving a progress payment from the owner.

- Owners are also becoming wise to the fact that liquidated damages limit their compensation for financial damages. As such, liquidated damage clauses are being excluded from contracts. This allows the owner to bring suit against contractors for actual financial damages.
- During the past, owners have lost many claims as a result of untimely turnaround of shop drawings and other submittal reviews. As such, a few contracts are being written putting the responsibility solely with the general contractor to ensure that shop drawings and other submittals are in accordance with the design documents. These contracts request copies of shop drawings and submittals to be transmitted to the owner and the owner's architect "For Information Only."

The risks stated and others which have not been included are onerous for a general contractor to overcome. These risks also tend to reduce a general contractor's profit margins.

As a result of the natural competitiveness of the construction industry, many contractors enter into a contract knowing the risks, but as if sprinkled with pixie dust, believe they are immune to the exposure. In order to reverse the trend of ever-increasing rates of failed general contractors, contractors must begin evaluating the risks involved in not only each project but the industry as a whole and shift the costs/losses from those risks to the party who controls the risk.

THE APPEAL (THE REWARDS)

Based upon a 1993 survey sponsored by this author of 111 general contractors, the rewards of the construction industry are as follows:
- The challenge and satisfaction involved with producing a physical reality from a concept on paper.
- The variety and diversity of projects and people associated with the construction industry.
- The personal rewards of a) making own hours, b) putting other people to work, and c) working outside.
- The financial rewards of getting a higher return on the initial investment than just keeping the money in the bank.
- The competitive nature of the business.

Chapter 2

ESTIMATING AND PROJECT MANAGEMENT ROLES

THE ESTIMATOR

Estimating is one of the most important aspects of a construction company's business operations. The lifeblood of the company is in its ability to secure new work profitably through either competitive bidding or competitive negotiation. Much of the credit for the success or failure of a construction enterprise can be attributed to the degree of capability of its estimating department and personnel. In an atmosphere of intense competition, such as exists in today's construction industry, the preparation of realistic, reliable, and accurate bids requires good judgment and estimating skills. Therefore, it is incumbent on any successful contracting organization to ensure its future through careful selection and training of estimators.

Assuming the work will be managed by competent field superintendents and project managers, the amount of profit attained on a given project is, to a large degree, reflected by the estimator's skills and experience. An estimator's skills enable him to organize the estimate and utilize the latest and most accurate techniques in preparing costs, while his experience enables him to visualize the construction of the project through its various stages.

The following requisites are essential in the making of a good estimator:

- Strong aptitude for mathematics through basic math (addition, subtraction, multiplication, division, decimal system), trigonometry and solid geometry (especially in the area of mensuration formulas).
- The ability to read and comprehend plans, addenda, specifications, special conditions, general conditions and other bid documents.
- The ability to mentally visualize the various phases of the construction project and price the job accordingly.
- Detailed knowledge of varying job conditions, latest construction techniques, productivity and availability of craftsmen, and capability and limitation of equipment.

- An intimate knowledge of costs for labor, equipment, material, transportation and other services.
- Capability to provide CPM schedules, budgets and cash flows on proposed projects.
- Experience in the kind of work being estimated or the ability to draw from the experience of others.
- Understanding of the mathematics of finance, risks and uncertainty. Strong in bidding strategies with a good "gut feel" for the numbers game.
- Knowledge in the field of construction bonding and insurance.
- Working knowledge of legal implications, including preparation of claims, requests for time extensions and rebutting claims by others.
- Familiarity with contracting with owners and subcontractors.
- Familiarity with functions of purchasing and expediting.
- Cognizance of capabilities and past history of competitors.
- Knowledge of the tools of the trade (i.e., computers, calculators, planimeter, scales, digitizers, plan measures, conversion tables, conversion factors, special forms, and productivity manuals).
- Access to and familiarity with information relating to materials, labor productivity, equipment productivity and costs of all kinds.
- Ability to formulate innovative ideas to solve special construction problems.
- Mediocre (as a minimum) design capability.
- An above-average amount of common sense.
- Ability to meet bid deadlines and still remain calm.
- Awareness of trade journals, service center reports, and other periodicals advertising projects to be bid.
- Intelligence, motivation, self-discipline, and honesty.
- Education through training and experience.
- Capability to use computers and the software available to the estimating functions.

Since it is unlikely that two projects will be constructed exactly alike and under the same conditions, the estimator must approach each new project with a fresh perspective.

THE PROJECT MANAGER

Project management is another important aspect of a construction company's business operations. A contractor will not remain solvent if profits are not realized on successfully bid/negotiated projects.

The project manager is accountable for the following functions.
- Maximize the profits on assigned projects.
- Complete each assigned project within the contractual time limit.
- Maintain good working relationships with owners, architects, engineers, subcontractors and suppliers.

The skills and attributes required of a project manager are listed below.
- Knowledge of construction scheduling techniques and applicability.
- Ability to communicate well (both verbally and written) with a wide range of personalities (i.e., owners, architects, engineers, city officials, government agencies, craftsmen, subcontractors, attorneys and accountants).
- Ability to read, interpret, and comprehend plans and specifications.
- Thorough understanding of local building codes.
- Working knowledge of construction and contract law.
- Capable of writing subcontract agreements.
- Capable of writing owner/contractor agreements.
- Ability to read potential crises situations and act appropriately.
- Courage to challenge an owner, architect, engineer, building inspector, subcontractor, vendor, etc., at the appropriate time.
- Expertise at recognizing and submitting valid claims.
- Competent negotiation skills.
- Above-average problem-solving skills.
- Experience in the types of work managing.
- Good organizational skills.
- Capable of writing intelligent, logical, non-emotional materials.
- Knowledge of construction methods and equipment.
- The following leadership skills and behaviors:
  - Integrity
  - Team building
  - Motivation
  - Interpersonal Relationships
  - Dependability

The project manager's roles and responsibilities include, but are not limited to the following:

- Study, review, and analyze the project documents including the plans, specifications, general terms and conditions, special terms and conditions and addenda.
- Analyze subcontractor and material supplier scopes of work to determine subcontract and purchase order awards.
- Write subcontracts and purchase orders.
- Write owner/contractor agreements, when applicable.
- Aid in selection of project team personnel (i.e., project superintendent, clerical personnel, craft foremen, etc.).
- Chair weekly job site meeting with subcontractors.
- Chair weekly job site meetings with owner and architect/engineer.
- Prepare project schedule and updates.
- Prepare and monitor submittal log.
- Act as contractor's agent for the project.
- Set up project files.
- Prepare and process monthly progress payment invoices.
- Approve and process monthly subcontractor and material supplier invoices.
- Price and negotiate change orders with owner.
- Review and negotiate change orders with subcontractors and material suppliers.
- Price and negotiate claims with owner.
- Review and negotiate claims with subcontractors and material suppliers.
- Analyze estimate to establish budget breakdown.
- Monitor project costs with respect to budget.
- Provide reporting to upper management with respect to project status (schedule and budget).
- Personally visit each project on a regular basis.
- Review submittals from subcontractors and suppliers.
- Prepare correspondence and respond to incoming correspondence as required.

For large and medium-sized contractors, a project manager is typically assigned to only one project and works on that project until its completion. Project managers for small construction companies typically are assigned three to eight projects to manage at any given time.

THE PROJECT SUPERINTENDENT

Coordination of the day-to-day activities of the physical project site is yet another key aspect of a construction company's business operations.

This responsibility falls into the hands of the project superintendent. The project superintendent is held accountable for the following functions:

- Assuring the quality of workmanship equals or exceeds the requirements of the drawings and specifications.
- Assuring that the working conditions of the project are maintained in a safe environment for workers, visitors, city officials, etc.
- Coordinating the daily work schedule of the project.

The skills and attributes required of a project superintendent are listed below.

- Experience in civil, architectural, and structural building construction systems.
- Working knowledge of mechanical, plumbing, electrical, fire protection, landscaping, and special systems.
- Ability to read, interpret, and understand construction plans and specifications.
- Knowledge of scheduling as a tool of the trade.
- Good verbal communication skills.
- Adequate problem identification skills.
- Working knowledge of local building codes.
- Ability to perform construction staking and use a construction level to develop grades.
- Intimate knowledge of construction techniques and methods.
- Ability to lead a diverse group of people

The duties, roles, and responsibilities of the project superintendent are listed below:

- Coordinate the daily activities of subcontractors and force account labor on the job site.
- Notify local building officials of building, electrical, mechanical, plumbing, fire protection, and other inspection requirements.
- Perform daily safety inspection of the project and implement corrective action as needed.

- Prepare documentation of the project in the form of daily logs, pictures, videos, and tape recordings.
- Perform daily inspection of work to ensure compliance to contract documents.
- Notify project manager of potential claim situations.
- Notify project manager of problems that have gone beyond the project superintendent's authorization.
- Prepare one-week and three-week "look ahead" schedules to be disseminated to all project subcontractors and suppliers.
- Coordinate staging of construction materials and equipment.
- Coordinate daily housekeeping chores.

Chapter 3

THE DECISION TO BID

CONTRACTING OPPORTUNITIES

Each day construction executives spend many hours researching various opportunities to perform their livelihood. There are many sources that list projects available for submitting bids. Among the services that advertise projects to be bid are those listed below:

- "F.W. Dodge Company Reports."
- "Associated General Contractors Reports."
- "Commerce Business Daily" published by the United States Government.
- Legal section of local newspapers.
- Private publishing firms.
- Trade magazines and journals.
- Local plan rooms.

The invitation to bid will normally include the following information:

- Name of project.
- Location of project.
- Name of architect.
- Name of engineers.
- Estimated dollar size of the project.
- Description of components of project.
- Anticipated date and time for receiving bids.
- Amount of deposit for bid documents.
- Number of bid documents available to each bidder (see Fig. 3.2).

Factors playing a role in determining which project to bid are listed below:

- Evaluating the risks of the project in comparison to the potential profit.
- Availability of qualified craftsmen and/or supervision.
- Number of construction companies that are likely to submit a bid.

- The dollar amount of uncompleted work the company currently has under contract. (i.e., amount of bonding credit)
- Profitability potential for the project.
- Experience in projects similar to one being advertised.
- Union or nonunion status of region.
- Geographic location of project.
- Owner's reputation.
- Architect's reputation.
- Complexity of project.

After upper management has selected a project to bid, an estimator is assigned to handle the estimating process.

AVAILABILITY OF BID DOCUMENTS

After reviewing the solicitation for bidders, the estimator must determine the following items with regards to bid documents:
- Number of sets available to a bidding party.
- Number of sets required for in-house use.
- Amount of deposit for the documents.
- Whether the deposit is refundable or non-refundable.
- Plan rooms that will have complete sets of documents.
- How subcontractors and suppliers will have access to the documents.
- Date that the documents may be picked.
- If the documents can be electronically retrieved and down-loaded.

This information is usually obtained by making a phone call to the architect/engineer, owner, or owner's representative for the project. The owner normally sets a limit on the number of sets of contract documents that the prime contractor may obtain. In private work, a deposit is generally required for each set of bid documents ordered. The deposit, which acts as a guarantee for the safe return of the documents, may range from $50.00 to $5,000.00 and is usually refundable.

For government projects the prime contractor <u>buys</u> the documents from the contracting agency. The cost may range from $10.00 for a one-half size set to $200.00 for a full-size set of drawings. (Note: There is an inherent danger using one-half size drawings in that the scale shown on the drawings is incorrect. The number obtained by "scaling" a measurement must be doubled to obtain the correct measurement.)

In order to be assured of a competitive price, a substantial number of quotations must be received from suppliers and subcontractors. Thus, the estimator must determine how those people may review the plans and specifications. In order to assure that there are no undue restrictions to subcontractors and suppliers the estimator should take the following steps:

- Order a few extra sets of the documents solely for the use of subcontractors and suppliers.
- Request that the contracting agency, architect/engineer, owner, etc. place at least two sets of plans in all plan rooms in the geographical region of the project (within 200-mile radius).
- Request that drawings and specifications be made available through electronic retrieval media.

PARTICIPATION OF THE SURETY COMPANY

At this time the surety company should be notified that the general contractor is anticipating preparing a bid for the project. The bonding company usually requires the following:

- Name, location, and estimated value of the project.
- Type and amount of bid security required.
- Date and time of bid opening.
- Whether or not a performance and payment bond is required.
- Time allocated by owner to complete the project.
- Type of work to be self-performed.
- Name of owner and architect/engineer.
- Value of uncompleted work on current projects.
- Amount for liquidated damages.

The estimator should complete an information sheet similar to Fig. 3.1 for presentation to the surety company.

Most owners require a bid security from each contractor presenting a proposal for a given project at the time of bid opening. This is commonly known as the bid bond. The bid security is a guarantee that the contractor will enter into the contract with the owner at the price indicated on the proposal documents. This security is addressed to the owner and signed and notarized by both the contractor and the contractor's surety company.

The bid bond may range from 1% to 20% of the value of the proposal and may be in the form of cash, certified check, or bid bond. If the contractor should refuse to enter into a contract for the stipulated amount, the owner could require the contractor to forfeit the amount specified on the bond. The cost of the bid bond is very nominal and is a general overhead item. However, many bonding companies do not charge the contractor for the service of providing a bid bond to those contractors whom they regularly furnish contract bonds.

Prior to providing the contractor with the bid security, the surety company will assess the contractor's ability to perform and finance the work.

With all other factors being satisfactory, the final gauge as to the amount of surety credit that will be extended to the contractor is his ability to finance contracts and absorb losses. Generally, his financing ability is based upon the net quick worth of the company and the line of credit available to the company through lending agencies.

The company's net worth consists of:
- Cash in the bank.
- Undisputed accounts receivable.
- Earned estimates due on uncompleted contracts.
- Corporate stock of quality companies.
- Other valid and collectible receivables.
- Miscellaneous investment assets.
- Liabilities such as accounts payable, stock holder's equity, promissory notes, etc.

The ratio of surety credit to net worth varies from one contractor to another and one surety company to another. Generally, the surety company will not allow the contractor to bid a project whose estimated value exceeds the contractor's net quick worth by no more than tenfold. (Thus it is important that the estimator be familiar with the maximum size of the job that the company will be allowed to bid in order to refrain from ordering bid documents on projects that the surety company will not bond.) The surety company will also limit the amount of uncompleted work on hand. The limit is usually no more than twenty times the contractor's net quick worth.

Except under unusual circumstances, the owner will require the qualified low bidder to provide a performance and payment bond. This is commonly known as a surety bond.

A surety bond is a three-party instrument between the contractor, obligee or owner (who is the party requiring protection), and the surety company. The surety company guarantees that the contractor will make whole any loss the owner might sustain by reason of the contractor's failure to carry out and perform all the conditions of the agreement entered into between the contractor and the owner. If the contractor fails or is in any way unable to fulfill his contractual obligations, the surety company then performs in accordance with the conditions of the bond. The surety company can then seek to recover its losses through the contractor, contractor's principals, and/or possibly even the owner.

The cost of the bond depends on the past performance of the contractor. In some states the best rate a contractor can be given is calculated by the following formula: $9.00 per thousand dollars of project value for the first $500,000; $6.50 per thousand dollars of contract value for the next $2,000,000; $5.40 per thousand dollars of contract value for the next $2,500,000; plus $4.50 per thousand dollars of contract value for the remaining value of the project.

An example of calculating the cost of a surety bond on a project follows:

EXAMPLE 3.1 Cost of Payment and Performance Bond

Given: a) Value of project is $5,000,000.
b) Contractor has "preferred" bond rate.

Calculations: a) $9.00 per $1,000 of value on first $500,000 .... $4,500
b) $6.50 per $1,000 of value on next $2,000,000 .. $13,000
c) $5.40 per $1,000 of value on next $2,500,000 .. $13,500

TOTAL COST OF BOND $31,000

The contractor should select the proper surety company and maintain a satisfactory business relationship with the firm chosen.

Upon approval of the bonding company for the contractor to submit a bid, the estimator can begin the task of preparing the estimate.

Figure 3.1 Project Information Sheet

NAME OF COMPANY: _____

ADDRESS OF COMPANY: _____
_____

PERSON SUBMITTING INFORMATION: _____

NAME OF PROJECT: _____
_____

LOCATION OF PROJECT: _____
_____

NAME OF OWNER: _____

NAME OF ARCHITECT/ENGINEER: _____

VALUE OF PROJECT: _____

TYPE OF BID SECURITY REQUIRED: _____

AMOUNT OF BID SECURITY REQUIRED: _____

IS A PERFORMANCE/PAYMENT BOND REQUIRED? _____

CALENDAR DAYS TO COMPLETE THE PROJECT? _____

PROJECT START DATE: _____

AMOUNT OF LIQUIDATED DAMAGES: _____

TYPE OF WORK TO BE SELF-PERFORMED: _____

VALUE OF UNCOMPLETED WORK ON CURRENT PROJECTS:
_____
_____

Figure 3.2 Invitation To Bid (Sample)

Sealed proposals are hereby requested by the public school district for construction of Jefferson Elementary School.

Date of Bid:	11/9/93
Time of Bid:	2:00 p.m.
Location of Bid:	Office of the Superintendent
	3 Main Street

Bid documents are available at the office of the superintendent upon receipt of a deposit of $100 per set. Each bidder may obtain three sets. Additional sets may be purchased for $150 per set. Deposit, in full, will be refunded to bona fide bidders upon return of full sets of documents.

A bid security in the amount of 5% of the bid will be required at the time of bidding.

The scope of work includes, but is not limited to, constructing a new 50,000 sq ft elementary school.

Please refer to the instructions to bidders for additional information.

Chapter 4

PREPARING FOR THE ESTIMATE

ENSURING SUBCONTRACTOR AND SUPPLIER COVERAGE

Upon receiving the bid documents the estimator should check the "Instructions to Bidders" section (see Fig. 4.1) of the specifications for the following items:
- Date of bid.
- Local time of bid.
- Location of bid opening.
- Whether or not a bid security is required.
- The type of subcontractors and materials that will be required.

In order to obtain a representative number of subcontractors and vendor quotations on bid day, it is important that the general contractor's participation as a bidder be well publicized. This is accomplished by notifying construction reporting services, advertising in regional newspapers, advertising in local trade journals, word-of-mouth, and requesting for sub-bids and material prices.

Requesting sub-bids and material prices may be accomplished by:
- Telephone notifications.
- Mailing out "bid request" postcards.
- Mailing out self-addressed returnable "bid-requested" letters.
- Faxing messages.

The format for the cards, letters, and faxes should identify the project, date, and time of the bid opening, the architect, and how bidding materials may be obtained. The solicitations should go out only to those subcontractors and suppliers whose normal scope of work is included in the technical specifications or noted on the drawings. An office file of subcontractors, suppliers, vendors, and specialty contractors should be maintained on index cards and computers. This file should be updated after each project that is bid.

BID DOCUMENT REVIEW

The next activity is for the estimator to become familiar with the plans and specifications. During this time the estimator should be getting an idea of the extent of the job to be estimated as well as the amount of detail shown on the drawings. This is accomplished by performing the following functions:

- Studying the site plan and orientation of the structure.
- Studying the floor plan for overall size.
- Studying the elevations for height and style of structure.
- Reviewing the sections and details for type of construction to be utilized.
- Reviewing the demolition, architectural, mechanical, plumbing, fire protection, electrical and other plans (food service, interior design, etc.) for an overview of those items.

The estimator should then thoroughly read the book of specifications. The specifications booklet will normally include the following:

- Instruction to bidders.
- Proposal forms.
- Form of owner-contractor agreement.
- General conditions.
- Supplementary general conditions.
- Special conditions.
- Summary of work.
- Alternates.
- Allowances.
- Technical specifications.

The estimator should have some note paper at hand while reading the specifications to jot down unusual requirements encountered and to note the items for which subcontractors and suppliers will be solicited.

A. Instruction to Bidders

The Instruction to Bidders is the portion of the document that relates the procedures to be utilized by all bidders in submitting a proposal. The instructions include the manner in which the bid is to be submitted; the date, time, and location of the bid submittal; how the envelope containing the bid is to read; the number of proposals to turn in; the amount of the bid security; additional information to be submitted with the bid, such as contractor's financial statement, project schedules, contractor's qualifications, schedule of values, minority business plan, etc.; and whether the bids will be read aloud or opened privately. See Fig. 4.1.

B. **Proposal Form**

The proposal form is the document utilized by the owner in evaluating the bids received and usually requires the following information:
- The lump-sum amount for the project.
- The number of days the price remains valid.
- Acknowledgment of receipt of all addenda.
- Prices for alternate methods or materials.
- Unit prices for various items.
- Occasionally, the owner may require the contractor to list the number of days required to complete the project.
- Signature of an officer in the company.
- Listing of subcontractors.

C. **Form of Owner-Contractor Agreement**

This section of the specification indicates the form of agreement between the contractor and the owner. This agreement may be one of the standard publications issued by groups such as the Associated General Contractors, American Institute of Architects, National Society of Professional Engineers, etc., or it may be a form developed by the owner. In any case, this document should be reviewed by the estimator with legal entrapments and risk shifting clauses in mind. If an article is unclear, clarification should be obtained from the owner prior to bidding the project. The terms and conditions in the agreement may be so unilateral for the owner that upper management may prefer not to bid that particular project.

D. **General Conditions of the Contract**

The general conditions spell out the rights and responsibilities of the parties involved. The most stringent demands are generally placed on the contractor since he has the responsibility for actually constructing the project.

The general conditions will usually contain the following subjects:
- Contract documents.
- Architect's responsibilities.
- Owner's responsibilities.
- Contractor's responsibilities.
- Subcontractors.
- Work by owner or by separate contractors.
- Miscellaneous provisions.

- Time.
- Payments and completion.
- Protection of persons and property.
- Insurance.
- Changes in work.
- Uncovering and correction of work.
- Termination of the contract.

E. **Supplementary General Conditions**

The supplementary general conditions add to or supplement the general conditions to meet the specific requirements of a particular project. Since the supplementary general conditions are tailored for the project, items that add cost to the general contractor may be unavoidable. Thus, it is necessary to thoroughly study the entire supplementary general conditions to note all items which must be covered in the bid and to decide what costs to allow for each item.

F. **Special Conditions**

Special conditions are incorporated by the owner to identify requirements that are also specific to the projects. Examples of special conditions topics are listed below:
- Job site safety.
- Employee conduct.
- Working hours.
- Project access.
- Security.
- Construction parking.
- Construction screening.
- Noise abatement.

The special conditions will normally affect both the general contractor's and subcontractor's cost of work.

G. **Summary of Work**

The Summary of Work briefly describes the activities which the contractor will be required to perform. It will also list the specific items that the contractor is not required to accomplish.

H. **Alternates**

In addition to the base bid, the owner may ask for prices on alternative materials, equipment or scope of work. These prices may be either adds or deducts to the base bid. This procedure is generally employed by the owner for the following reasons:

- Provides a means for making change orders prior to award of contract to alleviate "negotiating" the changes.
- Provides a means to eliminate items the owner really does not need in order to meet the owner's allocated budget.
- Allows the owner a choice of items on which to spend his money.

Lump-sum contracts are awarded on the basis of the base bid plus or minus any or all of the alternates requested. Thus, alternates deserve the same estimating care and consideration given to the base bid. Alternate bids involving only a few subcontractors or suppliers or the contractor's force account work can be processed easily on bid day. However, if the owner requires too many "minor" alternates or several "major" alternates which involve many or all of the sub-trades, there is typically insufficient time on bid day to evaluate all the combinations. The reasons for the shortage of time are inherent in the bidding process and are beyond the control of the best estimator. Therefore, the estimator should be conservative in analyzing alternate bids when time does not allow proper evaluation. On building work, it is common for alternates between the first and second bidder to vary by over 50% whereas the variance between the two contractors on the base bid is usually less than 5%.

One final note on alternates is that they are a necessary evil during the bid day process, and since the award of the project may hinge upon the outcome of the alternates, the estimator must have a good idea of the subcontractors and suppliers that will be affected. A serious error in judgment may cost the contractor the project (an even more serious error may cost the contractor dollars out of his profit margin or net worth).

I. **Allowances**

Allowances are prescribed sums of money specified by the Owner or Architect in the bid documents to be used in preparing certain aspects of the bid. They are used in order to 1) provide a quality basis where several products of different values might be available to the contractors; 2) provide a price for an item that has not yet been specified; and 3) allow the architect/engineer more time in designing a system that is incomplete when the bid package is released. They may be in the form of a unit price or a lump-sum. An example of a unit price allowance is $200.00 per thousand for brick or $10,000.00 for landscaping.

J. **Unit Prices**

In addition to the base bid and alternates, the Owner may request the bidding parties to submit unit prices for specific work activities. Unit prices may be either adds or deducts from the base bid or contract when associated with quantities. See the section on unit prices for more information.

K. **Technical Specifications**

The technical specifications are comprehensive presentations of the principal factors entering into the prosecution and completion of the work specified by the contract. They are generally organized in a manner that approximates the order in which the work will be accomplished at the construction site. These specifications include the type of material required and the workmanship that is expected. The material portions of the specifications usually refer to the physical properties, brand names and numbers, performance requirements, installation procedures, and handling and storage requirements. The Construction Specification Institute has developed a format which divides the major areas involved in building construction into sixteen major divisions. This format is widely accepted by architects, engineers, and contractors. The CSI major categories are listed below.

- General Requirements
- Site Work
- Concrete
- Masonry
- Metals
- Wood and Plastics
- Thermal and Moisture Protection
- Doors and Windows
- Finishes
- Specialties
- Equipment
- Furnishings
- Special Construction
- Conveying Systems
- Mechanical
- Electrical

The entire list including subcategories may be obtained through the Construction Specification Institute. The estimator or his delegate should perform a site investigation as the next task in preparing the bid.

SITE INVESTIGATION

A visit to the site by the principal estimator, following initial study of the tender documents, is advisable. In cases where time and/or distance interferes with the schedule for submitting the bid, a person other than an estimator may be used for the site trip provided a detailed report is prepared and presented to the principal estimator.

The notes prepared during preliminary study of the bid documents should be referred to during inspection of the site. During the site investigation, the investigator should orient himself in relation to the proposed work. The estimator should also attempt to visualize the manner and sequence in which the project will be constructed.

During the site investigation, the following information should be ascertained:
- Site access.
- Location and availability of electricity, water, telephone, sanitary sewer, and other utilities.
- Manner in which site will be drained during construction (if applicable).
- Location and availability of transportation facilities.
- Proximity of adjacent structures (to determine if protection or underpinning is required).
- A rough layout of the site locating proposed storage trailer and equipment locations.
- Apparent soil conditions.
- Local ordinances and regulations.
- Local building codes.
- Local building permit fees.
- Local labor complexion (union or nonunion).
- Labor rates for local craftsmen.
- Prices and delivery information of local material suppliers (concrete, gravel, lumber, etc.).
- The names of local subcontractors who will be pricing the project (if time allows, a visit should be made to those subcontractors).

- Conditions of roads and weight and height limits of bridges leading to the project.
- Availability of housing if project is away from normal working region.
- Banking facilities.
- Amount and type of security system required for project.
- Place and cost for dumping trash.

Site investigators should take adequate notes of their findings. Those notes play an important role in determining the final price that will be submitted. After the initial preparation and site visit, the estimator should be ready to begin quantifying the components of the project.

Figure 4.1 Instructions to Bidders (Sample)

Proposals

Sealed proposals for construction of the Jefferson Elementary School will be received by the public school district at the office of the superintendent until 3:00 p.m. on November 9, 1993. The sealed proposals will then be opened and read aloud.

The following documents must be included with the bid.
1. Bid Form
2. Subcontractor Listing
3. Qualification Form
4. Bid Security

The proposal envelope must include:
1. Name of Project
2. Name of Bidder

Bid Security

A bid security in the amount of 5% of the bid is required with each proposal. Approved forms of bid security are:
1. Bid Bond
2. Cashier's Check
3. Cash

Bid security for all but the two low bidders will be returned within fifteen days after receiving a notice of award of the contract. Should the successful bidder refuse to accept a contract for the project, that bidder's bid security shall be forfeited to the owner.

Performance Bond

A payment and a performance bond for the full amount of the bid will be required for the successful contractor.

Bid Forms

The bidder must complete each blank on the bid form.

Subcontractor Listing

Bidders must submit the list of proposed subcontractors with the proposal. The owner reserves the right to reject subcontractors at no additional cost.

Figure 4.1 Cont'd

Miscellaneous Information

Bidders must submit to the architect a schedule of values within twenty-four hours of the bid opening. Bidders must submit proposed labor rates for all trades for change order work to the architect within forty-eight hours of the bid opening.

Award of Bid

The owner has the right to reject any and all bids and to award the contract to the best evaluated bid (not necessarily the low bidder).

Validation Period

Each bidder must agree that the bid will not be withdrawn for a period of 90 days following the bid opening.

PREBID CONFERENCE

During the period of time between the invitation to bid and the day of the bid opening, the owner should elect to schedule a prebid conference and job walk.

The purpose of the conference is to allow the prime bidders, subcontractors and material suppliers to request clarifications of the bidding documents, to request substitutions of materials from those specified and to ask general questions of the architect, engineers, and/or owners concerning the plans, specifications, and terms and conditions.

The types of items and questions generally discussed in the prebid conference are listed below:

- Errors in the bid document drawings.
- Discrepancies between the drawings and specifications.
- Approval of materials other than those specified.
- Clarifications of details that are unclear.
- Project schedule.
- Potential problems.
- Discussion of contract terms and conditions.

In addition to the items relating to how the project is to be constructed and those mentioned above, the estimator may also determine the names and numbers of other prime bidders interested in submitting a bid and also the names of subcontractors and suppliers who are interested enough in the project to attend the prebid conference. Thus, the prebid conference can be an important part of putting the bid price together.

Chapter 5

PREPARING THE ESTIMATE

THE QUANTITY SURVEY

In estimating, the process of examining the bid documents to determine the amount of work required to build a project is defined as a quantity survey. This activity requires an understanding of geometric shapes, mathematics, and mensuration formulas.

The estimator's computations normally involve measurements of cubical contents, surface contents, and weights. Measurements will typically be stated in units of lineal feet (ln ft), square feet (sq ft), square yards (sq yd), units of 100 square feet (sqs), cubic feet (cu ft), cubic yards (cu yd), pounds (lb), tons (tns), and each (ea).

The estimator should prioritize the order in which quantities are to be surveyed. He should first determine the quantities of work that will definitely be performed by the general contractor's own forces. He should then quantify items which may not be covered by sub-trades for the specific project. Finally, if time allows, he should take-off miscellaneous items that will definitely be bid by subcontractors or vendors. This last step will allow the estimator to know a range of costs for a particular item. (An example of this last item would be a count of the plumbing fixtures. If there are 250 plumbing fixtures and the estimator knows from previous projects that a conservative subcontracted cost per fixture is $700.00, then he knows the cost for plumbing should be around $175,000.00.)

Accuracy is the keynote to good estimating. There are many "tricks of the trade" that can save time, reduce errors, and improve accuracy. However, there are no short cuts that should be taken at the expense of accuracy. The estimator must also be aware that there is a trade-off between the benefits of preparing an extremely accurate detailed quantity survey and the time and costs spent on such a detailed quantity take-off. There is a point where it becomes unfeasible to spend an inordinate amount of time on a particular detail as that extra time spent will not appreciably affect the accuracy of the take-off. To summarize, do not spend an hour taking off a fifty-cent item.

There are several rules that should be followed for quantity take-offs.

- Before starting to take-off quantities, examine the drawings. This will give the estimator an idea of where to locate the a) architectural plans, elevations, and details; b) structural plans, elevations, and details; c) site plans and details; d) electrical and mechanical plans and details; e) the number of stories in the building; f) the type of construction; and g) the layout of the building.
- Measure each item as it is shown. Do not arbitrarily round-off measurements, as this will probably add quantities to a particular item and thus costs to the estimate. Use dimensions as shown on the plans and elevations whenever possible. Do not scale unless absolutely necessary. The estimator should not make assumptions from the documents because he believes he knows more than the architect. It is better to ask the architect for a clarification.
- Mark the plans as an item is surveyed. An estimator will never be able to go through a quantity survey uninterrupted. Thus, marking the items as they are taken off will enable the estimator to begin at the point of interruption and will also reduce the number of items counted more than once or not counted at all.
- If the plans are not clear concerning an item, prepare a sketch to clarify the intent. Sketches can also be used to explain the input or how the estimator envisioned a particular construction technique.
- Make certain that the proper detail is used for each element. Be sure that each detail has been quantified. Once again marking the details as they are used will enable the estimator to spot one that has been missed.
- Segregate items that require special consideration. Do not mix items that have a different measurement or that will require different unit prices. Several examples are listed below:
    - Foundation walls should be separated from walls above grade.
    - Special coursings of brick should be separated from common coursing.
    - Hand-excavated activities should be separated from those that can be accomplished by machine.
    - Keyways should be taken off per lineal foot and not "thrown in" with wall forming.

- Duplex outlets should be separated from isolated outlets.

In most cases the quantity survey should be organized and accomplished in the order in which the estimator envisions the work to progress. For a building project, a suggested order of take-off is shown below:

1. Concrete and associated earthwork.
2. Masonry.
3. Miscellaneous metals.
4. Carpentry.
5. Doors and windows.
6. Interior finish items that the prime contractor plans to perform with his own forces.
7. Specialties such as toilet partitions, lockers, louvers, etc.
8. Equipment such as laboratory supplies, school equipment, refrigerators, etc.
9. Site work such as sidewalks, curbs, gutters, playground equipment, etc.
10. Alternates.
11. Sub-trade items for the purpose of having a "feel" for the prices on bid day, (i.e., mass excavation and backfill, fire protection, HVAC, electrical, stucco, painting, acoustical ceilings, etc.).

The order of the take-off should not be varied except for special projects. The purpose of establishing and maintaining a methodical approach is to order one's thinking in an almost mechanical pattern.

There are several "tricks of the trade" and mathematical shortcuts that can be utilized in the quantity survey phase.

- Abbreviate whenever possible (use standard abbreviations).
- Keep all dimensions, figures, and areas that might be useful later close at hand. An example of this is the length and depth of the perimeter basement wall. This figure can be used to calculate 1) formwork, 2) concrete, 3) footing drain, 4) waterproofing, 5) backfill, 6) concrete footings, and 7) point and patching.
- Use symmetry whenever possible.
- Combine measurements for items that are exactly alike but may be shown in several different locations. An example of this would be the coping on top of a parapet wall. This is performed easily by maintaining "collection sheets" for those pieces of work.

- Take advantage of duplications in design other than symmetry. As an example, floors two through six of a building may be identical, thus allowing one take-off and a simple multiplication of five to calculate all the quantities.

As each item is taken-off, the dimensions or units are entered onto collection sheets. The purpose of these sheets is to collect items that are similar in order to simplify the take-off sheets, save time, and reduce the possibility of error. After the extensions of the units have been made, the results are transferred to pricing sheets.

The estimator must be mindful of construction techniques during the quantity survey so that pricing of labor and equipment can be accomplished quickly and easily, without extra time-consuming work. As an example, the quantity of concrete to be placed direct should be separated from that to be placed by pumping.

The estimator must also consider minimum order, waste, and dimensional variances in performing quantity surveys. Examples include the following:
- The amount of drywall needed for a wall 10 ft long by 7.5 ft high is three sheets (4 ft x 8 ft) which is 96 sq ft rather than 75 sq ft.
- The amount of carpet needed for a 9 ft x 10 ft room is 12 sq yd rather than 10 sq yd since carpet is bought in 12 ft wide rolls.
- For new walls attaching to old walls, painting should be extended to the existing corner on each side of the wall.
- A job that has one room that is 10 ft x 10 ft requires 135 sq ft of floor tile (rather than 100 sq ft) since tile is bought in boxes containing 45 sq ft.

The quantity survey should now be summarized into the various units of work needed to construct the project. The purpose of the entire quantity survey and summarization sheets is to facilitate pricing of the labor, material, and equipment costs, and to provide the estimator with an "idea" of the cost of certain subcontractor items of work.

ESTIMATOR'S DATABASE

In order to assure that estimates are produced in a uniform and consistent manner, the estimator must maintain a reliable database from which to extract labor productivity factors, labor rates, equipment costs, insurance costs, tax rates, subcontractor unit prices, and material costs.

For estimating without the use of computer software, a manual system should be devised using an organized method to retrieve the information.

The numbering system used by the Construction Specification Institute is the preferred set-up for a database. An example of a manual database is shown in Fig. 5.1.

Most of the estimating software packages marketed nowadays require an independent database to be used in the pricing of an estimate. The information used in an estimating software database can be obtained from an estimator's hand written and manually produced database. Commercial databases are also available for certain software systems.

The estimator must update the database on a regular basis. One of the best methods of updating and augmenting a database involves reviewing each bid received on a project and thus changing existing items or adding new items to the database.

A few trades (mechanical, plumbing, and electrical) publish material costs on a regular basis (monthly, bimonthly, etc.). Incorporating those trade material costs into a computer software program will allow the trade estimator to maintain a current database at all times.

Figure 5.1 Manual Database

DESCRIPTION OF ITEM: _____

_____

_____

_____

CSI CODE NUMBER: _____

MATERIAL COST: _____

APPLICATION RATE: _____

WASTE FACTOR MATERIAL COST PER UNIT: _____

SOURCE OF MATERIAL COST: _____

COMPOSITE CREW: _____

_____

_____

_____

COMPOSITE CREW COSTS: _____

COMPOSITE CREW RATE: _____

LABOR PRODUCTIVITY: _____

SOURCE OF LABOR PRODUCTIVITY: _____

SUBCONTRACTOR UNIT COST: _____

_____

SOURCE OF SUBCONTRACTOR UNIT COST: _____

_____

MISCELLANEOUS COSTS: _____

_____

_____

DATE OF LATEST INFORMATION: _____

LABOR RATES

In order to produce a labor estimate for an activity, the estimator must be familiar with components of the labor rate. Labor costs are measured in dollars per hour. A labor rate is produced by combining the items listed below (see Table 5.1):

- **Base wage rate**

    The base wage rate is the amount of money earned by a craftsman for each hour worked. The base wage rate is determined using the following factors:
    - For union contractors, the local wages mutually agreed upon by the Trade Unions and the contractor or contractor's industry association.
    - For nonunion contractors, the criteria listed below affect the base wage earned by a construction worker:
        - Type of work.
        - Experience of worker.
        - Employment posture of area.
        - Whether or not the project is governed by Davis-Bacon wages.

- **Union Employee Benefits**

    Union benefits include items such as medical, dental and vision insurance, vacation pay, pension, training, industry advancement, dues, etc., which are paid directly to the union. The union benefits are applied to each hour that the union employee works for the employer. These benefits are negotiated between the union and contractor or contractor's industry association.

- **Nonunion Employee Benefits**

    Nonunion employers have the benefit of deciding what type of benefits, if any, to provide employees. The nonunion employer calculates the cost of the benefits on a monthly basis and then divides by either 173 (52 wk x 40 hr/wk / 12 months) or 160 (4 wk/mo @ 40 hr/wk) to obtain a cost per hour to apply to the base wage rate.

- **Federal Insurance Contributions Act (FICA)**

    The federal government requires employers to pay a tax for the purpose of providing retirement benefits. The act has two components. The first is FICA, which for 1995 was 6.45% of the first $61,200 earned by employees. The second is Medicare, which is

1.2% of the first $120,000 earned by employees. Both the employer and employee are required to contribute the same amount.

- **Federal Unemployment Insurance Tax (FUTA)**

This is a tax collected by the federal government for the purpose of providing funds to compensate employees who have been laid off. The current cost is 0.8% for the first $7,000 earned by an employee in a calendar year. Thus the maximum tax per employee is $56 per year.

- **Worker's Compensation Insurance**

This insurance provides benefits to employees of the contractor in the event of injury or death arising out of an accident which might occur while the employee is working on or in association with a construction project. The amount of the benefits paid are specified by various state and federal compensation laws.

The premium for worker's compensation insurance is calculated by multiplying the amount of money a worker earns by a factor which is based on the type and classification of work performed by the craftsman.

The premium base is multiplied by the contractor's experience modification factor. This factor is computed by analyzing the contractor's safety record and amount of compensation provided to the contractor's employees over a three-year period. If the contractor has a poor safety record during any fiscal year, it takes three years to overcome the increased worker's compensation premiums.

In addition to providing the benefits required by law, the standard worker's compensation policy affords employer's liability insurance against claims for bodily injury or disease arising out of employment. See Table 5.2 for a sampling of various worker compensation rates for different states.

An example of calculating the costs for worker's compensation insurance is shown below:

EXAMPLE 5.1 Worker's Compensation Cost Calculation

> For a masonry subcontractor working on a project in California with an experience modification factor of 0.82, the worker's compensation premium is $11.931 per $100 of straight time payroll (14.55 times 0.82). (Some states require the premium to be applied to the entire payroll, including premium for overtime.) If the masonry subcontractor had a payroll of $100,000 for a certain project the premium would be $11,931.

- **State Unemployment Insurance Tax (SUTA)**

    This is a tax collected by state governments for the purpose of providing funds to compensate employees who have been laid off. The rate varies by state. Many states have a standard rate which is then modified based upon the employees lay-off record. As with the Federal Unemployment Insurance Tax, the rate is applied only to the first $7,000 earned by an employee in a year.

    For example, the rate for first time employees in California is 3.4%. Thus, first-time employers pay a maximum of $239.90 per employee per year.

    An interesting note about both the Federal and State Unemployment Insurance Tax is that employers with permanent or semi-permanent employees can use FUTA and SUTA to either increase profits or increase competitiveness. For example, if an employer has 250 employees through the year with an average wage of $20,000 per employee, the maximum for unemployment taxes to be paid is $73,975 ($295.90 times 250). Applying the full unemployment taxes rate (0.8% and 3.4%) to the wages yields $210,000. Thus, if the contractor is successful in his bid efforts using the full rate, he will sustain a windfall profit of over $120,000 in the scenario. The contractor may elect to reduce the labor rate by calculating his actual unemployment rate. For this contractor the actual rate is 1.48% ($295.90 divided by $20,000).

Table 5.1 Labor Rate Calculations

UNION CONTRACTOR CRAFTSMAN
Trade - Electrician, Journeyman

A.	Base wage rate[1]	$24.30
B.	Health and welfare[1]	2.80
C.	Training[1]	0.05
D.	Pension[1]	2.00
E.	Vacation and holiday[1]	1.65
F.	FICA (7.65% of A)[2]	1.99
G.	State Disability Insurance (.9% of A)[2]	0.23
H.	Federal Unemployment Insurance (.8% of A)[2]	0.21
I.	State Unemployment Insurance (3.4% of A)[2]	0.88
J.	Workmen's Compensation Insurance (7.65% of A)[2]	1.86
K.	Liability Insurance (3% of A)[3]	0.73
	TOTAL HOURLY BURDENED WAGE RATE	$36.70

Note: (1) From union labor agreement.
(2) From government requirements.
(3) From insurance carrier.

NONUNION CONTRACTOR - CRAFTSMAN
Trade - Electrician

A.	Base Wage Rate	$12.00
B.	Medical Insurance[1]	0.00
C.	Vision Insurance[1]	0.00
D.	Dental Insurance[1]	0.00
E.	Pension[2]	0.00
F.	Vacation[2]	0.00
G.	Holiday[2]	0.00
H.	FICA (7.65% of A)[3]	0.92
I.	State Disability Insurance (.9% of A)[3]	0.11
J.	Federal Unemployment Insurance (.8% of A)[3]	0.10
K.	State Unemployment Insurance (3.4% of A)[3]	0.41
L.	Workmen's Compensation Insurance (9% of A)[3]	1.08
M.	Liability Insurance (3% of A)[4]	0.36
	TOTAL HOURLY BURDENED WAGE RATE	$14.98

Note: (1) Nonunion employers may make these benefits available for workers to pay out of their own wages.
(2) Nonunion employers do not typically provide this benefit.
(3) From government requirements.
(4) From insurance carrier.

Table 5.1 cont'd

UNION OR NONUNION CONTRACTOR - SUPERINTENDENT

A.	Base Wage Rate	$20.00
B.	Medical Insurance[1]	2.00
C.	Vision Insurance[1]	0.50
D.	Dental Insurance[1]	0.50
E.	Pension[2]	2.00
F.	Profit Sharing[2]	2.00
G.	Holiday[3]	0.75
H.	Vacation[4]	0.83
I.	Sick Time[5]	0.83
J.	FICA (7.65% of A+G+H+I)[6]	1.71
K.	State Disability Insurance (.9% of A+G+H+I)[6]	0.20
L.	Federal Unemployment Insurance (.8% of A+G+H+I)[6]	0.18
M.	State Unemployment Insurance (3.4% of A+G+H+I)[6]	0.76
N.	Workmen's Compensation Insurance (4% of A+G+H+I)[6]	0.90
O.	Liability Insurance (2% of A+G+H+I)[7]	0.45
	TOTAL HOURLY BURDENED WAGE RATE	$33.61

Note:
(1) Monthly premium divided by 173.
(2) Percent of A (varies with each contractor).
(3) Seventy-two hours of holiday times hourly rate; divide amount by 1928 (working hours in a year) (2080 hours minus 72 hours holiday time minus 80 hours vacation).
(4) Eighty hours of vacation times the hourly rate divided by 1928.
(5) Eighty hours of sick pay times hourly rate divided by 1928.
(6) From government requirements.
(7) From insurance carrier.

The table for calculating hourly wage rates can be set-up on computer spreadsheet software. This will allow updates and corrections to be made easily.

Table 5.2 Worker's Compensation Rates for Selected Trades and States

State	General Carpentry	Masonry	Steel Erection
Arizona	25.02	18.00	19.23
California	24.00	14.55	29.13
Florida	22.70	20.16	36.02
Kansas	9.03	9.60	13.93
New York	14.08	15.16	23.39
Wyoming	5.86	5.86	5.86
National Average	17.65	14.87	37.36

PRICING LABOR

In the construction industry, labor is the manpower required to perform the activities of work inherent in a project. The cost of labor in a construction project will be a large percentage of the total cost for that project. As such, the estimator must realize the importance of carefully and accurately pricing the labor cost of each individual activity. A bad labor estimate may have the following adverse affects:

- Cause the contractor to lose the project.
- Cause the contractor to lose money on the project if he or she is the low bidder.
- Cause the contractor to go out of business.

The most common method used to determine the labor units required to perform the work is using production rates applied to take-off quantities. A production rate is defined as the amount of work produced by a person, crew of people, or piece of equipment in a specified period of time, usually either an hour or a day. Examples of production rates are shown below:

- A carpenter can install one door per hour.
- A concrete crew can place 350 cu yd per day.
- A 3/4 yd backhoe can excavate 10 cu yd per hour.

The main source for the production rates is trade publications such as "Means Cost Data," "Richardson Construction Estimating Standards," "Walker's Building Estimator's Reference Book," and many others. These publications list the activity, the crew required to construct an activity, and the amount of work the crew can accomplish in a day. Another source of production rates available to estimators is the job cost history files from the company's accounting files. These files provide information that indicates actual production rates required to perform work.

When using production rates from published sources, the estimator must keep in mind the following:

- The production rates given are based on industry averages.
- The production rates could be higher or lower based on project conditions, location, availability, and use of construction equipment, quantity of work to be performed, and experience of the contractor.
- Rates are not available for every item of the take-off (in this case the estimator's judgment and crew-size calculations must be used).

The following steps should be taken in pricing labor using production rates:
1. Determine the base wage rates that apply to each craft for the project.
2. Determine the crew of craftsmen required to perform the work.
3. Calculate an hourly rate for the crew.
4. Choose the table which comes closest to conditions anticipated during quantity survey.
5. Multiply the production figure by the composite rate to arrive at a unit cost.
6. Adjust the unit cost to compensate for job site conditions and small quantities of work. For example, if an item of work has a small quantity, the unit cost should be increased to compensate for the lack of production.

Consider the following example: The estimator has determined that a project has 500 cu yd of wall concrete that should be placed by crane and bucket. During the site investigation the estimator determined that the following burdened wage rates apply to the area in which the project is located.

EXAMPLE 5.2 Composite Crew Cost per Hour and Resulting Unit Price

TRADE	WAGE RATE/HR
Cement Finisher	$18.00
Laborer	$16.00
Crane Operator	$19.00

He then determines his crew size and the crew hourly rate.

1. Cement finisher (1) at $18.00/hr.   = $ 18.00
2. Laborers (3) at $16.00/hr.   = $ 48.00
3. Crane Operator (1) at $19.00/hr.   = $ 19.00
4. Foreman (1) at $20.00/hr.   = $ 20.00
                                                            $105.00

For an eight-hour day the crew rate is $105.00 x 8 = $840.00

According to "Engelman's General Construction Cost Guide", the crew can place 60 cu yd of wall concrete by crane and bucket in one day.

Thus the unit cost to place wall concrete by crane and bucket becomes $840.00 ÷ 60/cu yd = $14/cu yd.

Next, the estimator must adjust for actual conditions he believes will be at the job site. During his quantity survey and while studying the drawings, he noted that for this job the crane and operator would be sitting on an embankment, unable to see the wall being poured. Therefore, a flagman must be added to the crew and the unit cost adjusted accordingly.

One day for an ironworker (normally used as a flagman) is $18.00/hr x 8 hours =$144.00. This will add 144 ÷ 60/cu yd = $2.40/cu yd to the unit cost. Thus the unit cost for labor to place the wall concrete with crane and bucket becomes $16.40 per cubic yard.

Crew-size calculations are also used on situations where production figures are not available. In this case the estimator must visualize the crew and the amount of time required to completely perform an activity.

Consider the following example: During the quantity survey, the estimator has determined that there are 15 guard posts in the project. The posts are 8 ft long made of 6 in. pipe. They are 4 ft in the ground with a 2 ft concrete base and then the pipe is filled with concrete. The estimator figures one laborer with a foreman part of the time can handle this activity.

EXAMPLE 5.3 Labor Estimate of an Activity Using Crews

	CREW	HOURLY RATE
1.	Laborer (1) at $16.00/hr	$16.00
2.	Foreman (1) at 6 min/hr	$17.00 x 6/60 = $1.70
	HOURLY RATE	$17.70

	ACTIVITY	TIME DURATION
1.	Receive and unload pipe	.2 hour
2.	Layout location of guard	.4 hour
3.	Dig hole for concrete	.6 hour
4.	Pour concrete in hole	.2 hour
5.	Set guard post and tie it off	.3 hour
6.	Fill guard post with concrete	.5 hour
	TOTAL TIME	2.2 hours

Total unit cost for installing one guard post is: $17.70/hr x 2.2 hrs = $38.94. Since the worker will not be 100% efficient and will have other things to do (drink of water, go to portable toilet, and so forth), a productivity percentage must be applied. Normally a fifty-minute hour can be used; therefore, the unit cost to be used in the estimate is $38.94 x 60/50 = $46.73 each.

The estimator should also consider the quantity of work before finalizing the unit price. Typically, if the quantity of work increases, the unit price decreases.

As stated above some estimators and companies use experience that has been gained on past projects to arrive at unit costs for labor.

A solid cost control program can be priceless to the estimator. With a little study, the estimator can take a completed project and arrive at competitive unit cost figures to be used in the future for the appropriate projects. This study consists mainly of analyzing actual costs versus estimated costs and applying a factor for increased wages and/or tougher working conditions.

During the discussion of pricing labor, adjusting the labor figure has been mentioned several times. Although adjusting the labor for job conditions is mainly a judgment call by the estimator, there are several factors that he can analyze for use in making his decision. These factors are discussed below.

Quantity of Work

The estimator must determine if an item has sufficient quantity to allow the work to be completed within the parameters established. If the quantities are low, the price of labor on a unit basis will be higher. If the quantity is high, there may be some economies overlooked by the estimator.

General Economy

This applies to the geographical area in which the project is to be built. Items that should be reviewed and evaluated under this category are:

- The business conditions in area of project.
- The construction volume in area of project.
- The employment situation in area of project.

Construction productivity tends to decline as the general economy and employment picture improves. This is due to the fact that the best supervisors and craftsman will more than likely have a job at any time, whether the economy is bad or good. As the economy and thus the construction business outlook gets better, the contractor has fewer experienced and qualified craftsmen to draw from.

An example of this is that during good economic conditions, union halls may be empty with contractors still requiring craftsmen. The unions will solicit inexperienced people to fill the requirements and issue those people a permit to work as a union member.

Project Supervisor

Normally, the estimator will know who the superintendent will be if the company is the successful low bidder on a project. If the superintendent has a record of besting his labor budget, the estimator might want to take advantage of that. Otherwise, the estimator should not consider this category as highly as some of the others.

Labor Conditions

A check during site investigation should be made to determine if experienced craftsmen are available locally. If they are not available, the contractor will have to pay a premium to bring in the qualified craftsmen.

Weather

The estimator must check past weather conditions for the area in which the project is located and the attitude of the craftsmen toward the weather. For example, craftsmen in Missouri have been known to leave the job site at the first raindrop, while craftsmen in Washington will work right through a torrential downpour.

Union Craftsmen

The estimator must also be familiar with work rules imposed by the Building Trades if union craftsmen are to be used. For example, the union contract might require an oiler for a tower crane. A tower crane does not need an oiler, but the extra cost must be added. Another example is that the contract might require the contractor to provide 20 minutes to allow workmen to put up tools and walk to their cars. This would require the estimator to add 4.2% to the labor column for lost production.

Escalation

If the project duration is longer than one year the estimator might have to add for an escalation in labor for activities to be performed later in the life of the project. After the labor costs for an item are calculated from the pricing sheets, the costs are transferred to the applicable line item of the main summary sheet, which is addressed elsewhere.

The final step in figuring the costs of labor for a project is to calculate the payroll burden. Payroll burden consists of payroll taxes, payroll insurance and fringe benefits. These items are discussed elsewhere in this book.

Overtime

If the project has scheduled overtime as part of the requirements, the effect of the overtime on productivity, as well as the premium cost, must be considered. The Mechanical Contractors Association of America (MCAA) <u>Labor Estimating Manual</u> recommends multiplying labor man-hours by inefficiency factors ranging from 1.12 for a five-day week, nine hours per day schedule to 1.77 for a seven-day week, twelve hours per day schedule. For the seven-day week, twelve hours per day schedule the effective time worked is 47.5 hours per worker versus the actual work time of eighty-four hours.

Shift Work

The MCAA <u>Labor Estimating Manual</u> recommends increasing man-hours for second-shift work by a factor of 20% and third-shift work by a factor of 30%.

Other Factors

Other factors which are rarely experienced on most projects, but do affect labor productivity and must not be overlooked are described below:

- <u>Stacking of trades</u> - Operations taking place within physically restraining conditions resulting in a congestion of personnel.
- <u>Morale and attitude</u> - Operations taking place in either a physically hazardous environment or a mentally disturbing atmosphere (excessive changes, over-inspection, bird dogging, etc.)
- <u>Concurrent operations</u> - Stacking of trades into a compressed schedule and/or work environment.
- <u>Beneficial occupancy</u> - Operation of trying to complete a project with the building attempting to be used for its intended purpose.

MATERIAL COSTS

Material in the construction industry can be described as the elements of which a construction project is composed. This includes the concrete, steel, doors, windows, paint, drywall, air conditioning equipment, lights, wiring, piping, plumbing fixtures, etc. However, the only material the estimator is concerned with is that which will be purchased directly by the general contractor.

Material prices will be quoted in two different forms.

The first form is on a unit price basis. The material supplier will phone or send a material list to the estimator indicating prices for a unit quantity. The estimator has the responsibility (during the quantity survey phase) of calculating the quantity of each item required for the project and then extending the quantity (with a waste factor) by the material price to arrive at a total cost for that item. For example, Acme Brick Company phones in a price of $150.00 per thousand for standard size face brick. The estimator must multiply that cost by the number (including waste) of standard size face bricks that he calculated would be required for the project.

The second form is on a lump sum basis. The material supplier will perform his own quantity survey and submit his scope of work and a lump sum price to the estimator. The estimator must still familiarize himself with the item so he can intelligently compare different supplier's scopes of work to ensure that the low bidder on an item has included all the material needed for a complete installation. (Many structural steel fabricators exclude the cost for expansion joint covers in their scope.) Examples of this would be reinforcing steel, cabinetwork, doors, windows, and miscellaneous specialties such as toilet accessories.

During pricing of the material one item that is easily overlooked is freight charges. The estimator must either verify that the price includes freight to the job or add the cost of freight to the bid.

As the pricing of material is completed or received, those prices should be entered onto the applicable line of the main summary sheet.

The final activity in pricing the material is to calculate sales tax on the total material costs required for the project. Sales tax is discussed elsewhere.

One other item of importance concerning pricing of material concerns "discounts." Suppliers who quote on a unit price or item (concrete, block, brick, lumber, and so forth) will usually provide a unit price quotation less a discount if payment is made by a certain day. As an example a concrete supplier may quote $40.00 per cubic yard less 5% if paid by May 10. In this case the estimator has a range of prices to use for the concrete material. The least amount per cubic yard is $38.00 ($40.00 less 5% discount). The greatest amount is $40.00. The estimator should consult with upper management to decide which price to use.

EQUIPMENT PRICING

Developing the costs for the equipment to be used on a specific project involves the following activities:

- Choose the proper type of equipment.
- Judge the amount of time the equipment will be needed on the job site and thus calculate the cost of the equipment.
- Figure the cost of fuel for the equipment.
- Figure the cost of maintenance.
- Figure the cost to move the equipment to and from the job site.

The first step in estimating equipment costs is calculating hourly costs for specific equipment. During the quantity survey, the estimator should be developing an idea of the equipment needed for the project. He should make a list of the equipment for pricing after the quantity survey is complete.

A note of importance is that the estimator's work in this area is theoretical. His choice of equipment may not match what is actually used but there should be a logical similarity and an equivalence of cost.

After the quantities of work have been calculated, the estimator can apply established production rates (the rates may be from equipment manufacturer's performance guides, company records, or estimating guidelines) to the amount of work to be performed to establish the amount of working time a machine will be used. This is the minimum amount of time a piece of equipment will be on the job site. This time will also be used to calculate the on-site operational costs. These may be obtained from equipment manufacturers, the "Equipment Green Book," or company records. Operational costs include the costs for fuel, lubrication, repairs, and mechanic's time.

After developing the amount of time a piece of equipment will be on the job site, the estimator must determine the rental rate to apply. Rental rates are usually quoted in dollars per hour, dollars per day, dollars per week, and dollars per month. The estimator should figure the costs in all four ways to determine the most economical. The rental rate for a piece of equipment includes the costs of insurance and depreciation. If the company owns the equipment, upper management should be consulted for the correct rate (in recession periods, management may determine to apply no rental rate). If the company does not own the equipment, the estimator should consult several equipment rental agencies to find the lowest rental rate. (The estimator should add extra insurance, sales tax, and repair costs for rented equipment.)

The next step in pricing equipment is to determine the costs for general types of equipment needed for the project. This includes pick-up trucks, flat-bed trucks, fork-lifts, scissor-lifts, cranes, etc. This equipment is generally priced by the number of months the piece will be used on the job site.

Rental rates are obtained from in-house management or equipment rental agencies. Operating costs are figured from estimated daily hours of operation or historical data.

The final step in calculating equipment costs is to figure the mobilization, set-up, and demobilization costs.

Two examples of pricing the equipment follow:

Example 5.4 Calculating the Cost for Specific Use Equipment

Cubic Yards of Footing Excavation: 350 cu yd
Equipment: Backhoe with 1/2 cu yd Bucket
Production Rate (from company data): 20 cu yd/hr
Rental Rates: $15.00/hr
$100.00/day
$400.00/week
$2,000.00/month
Operational Costs: $6.00/hr
Miles to Project: 50 mi
Weight of Backhoe: 5,000 lb
Commercial Transportation Rate: $.25/100 lb/mi

A. Calculate the number of operational hours

350 cu yd ÷ 20 cu yd = 17.5 hr. Applying a factor of 45 minutes actual work per hour provides 17.5 ÷ 45/60 = 23 1/3 hours of operation; thus the machine will be operational for three days. However, the estimator must still use his judgment in determining the total length of time the backhoe will be on the job site. In this case he assumes 35 cu yd of concrete for the footings will be poured each day. Thus the minimum amount of days the backhoe will be on the job site is 350 ÷ 35 = 10 days. Allowing for one weather day, one day of mobilization, one day to get out in front of the concrete crew, and one day of demobilization will give 14 working days at the job site. The estimator can now figure the cost of the equipment.

TYPE OF RENTAL	RATE	TIME	COST
Hourly	$ 15	112 hr	$1,660
Daily	$ 100	13 days	$1,300
Weekly	$ 400	3 wk	$1,200
Monthly	$2000	1 mo	$2,000

Thus the equipment rental should be estimated at $1,200.00 for the project. Operational cost is 23 1/3 hours x $6/hr = $140.00.

In calculating the mobilization and demobilization costs, the estimator should consider both commercial and company transportation. Since both mobilization and demobilization costs are equal, the estimator figures the cost for one way and multiplies it by two.

Cost of commercial transportation:
    A. 5,000 lb x 25¢/100 lb/mi x 50 mi x 2 trips = $1,250.00

Cost of company transportation:
    A. Assume two hours round trip for truck driver
        2 hr x 2 x 2 trips x $10.00 x 1.20(labor burden)= $96.00
    B. Two-hour round trip for company semi and trailer
        2 hr x 2 x 2 trips x $15.00 = $120.00
    C. Operational costs for semi and trailer
        2 x 2 x 2 x $10.00 = $ 80.00

Total costs for company to mobilize and demobilize = $296.00

Thus the mobilization and demobilization charges to be included for the backhoe are $296.00.

In summarization, the amount of money to be included in the estimate for the backhoe is shown below:

    Rental of Backhoe   = $ 1,100
    Operational Costs of Backhoe   =   140
    Mobilization and Demobilization Charges   =   296

    TOTAL $ 1,536

EXAMPLE 5.5 Calculating the Cost for General Use Equipment

Duration of Project: 12 mo

Equipment: Pick-up truck for superintendent (company owned)

Operational Costs: $3.75/hr

Rental Rate: $250/mo

Miles to Project: 50

    A. Operation Costs

From past projects, it has been determined that a superintendent's pick-up is in actual use 4 1/2 hours per day. Thus the operational costs can be calculated by performing the following calculation:

4 1/2 hr/day x 22 days/mo x 12 mo x $3.75/hr =$4,455.00

B. Rental Rate of Equipment
$250.00 x 12 mo =   3,000.00

C. Mobilization and Demobilization Costs
50 mi x 50 mi/hr x $3.75/hr x 2 trips =   7.50

TOTAL COSTS $7,462.50

In this case, since the company owns the pick-up, upper management may decide to either exclude item B altogether from the bid price or include a percentage of item B in the bid price. However, the operational, mobilization, and demobilization costs are true expenses that must be captured in the overall estimate.

UNIT PRICES

The purpose for unit prices is the same as for alternates. Whenever possible, the contractor should submit separate prices for additive and deductive unit prices. Deductive unit prices are typically priced lower than additive unit prices. This is due to the fact that unit prices are given higher mark-ups than base bid work. As such, if the same price is submitted for both add and deduct unit prices, the contractor may be faced with a reduction of base bid profit margin when an owner facilitates a change order based on a deductive unit price. As a note, many subcontractors will submit additive and deductive unit pricing for the same reason.

An example of unit pricing and the potential effect of deductive unit pricing is shown below.

EXAMPLE 5.6 Unit Price Calculation

The owner has asked for unit prices for the following:

- **4" Slab-on-grade (SOG) with 6 x 6 6/6 wire mesh on a 4" gravel base with 6 mil polyethylene vapor barrier (sq ft basis).**
- **6" Slab-on-grade with 6 x 6 6/6 wire mesh on a 6" gravel base with 6 mil polyethylene vapor barrier (sq ft basis).**

A concrete subcontractor has supplied the general contractor with the unit prices shown below.

Activity	Add Unit Price	Deductive Unit Price
4" SOG	$2.00	$1.80
6" SOG	$3.00	$2.70

For additive unit prices, the general contractor should use the following mark-up categories:
- General conditions
- Overhead and profit
- Liability insurance
- Builder's risk insurance
- Payment and performance bond

The deductive unit prices should be submitted as the subcontractor quoted, with no mark-ups applied. The additive unit prices to be submitted to the owner are shown below.

Item	4" SOG	6" SOG
Sub Unit Price	$2.00	$3.00
General Conditions (5%)	.10	.15
OH&P (10%)	.21	.32
Liability Insurance (2%)	.04	.06
Builder's Risk Insurance (1%)	.03	.04
Payment & Performance Bond (1%)	.02	.04
Total Unit Price	$2.40	$3.61

Thus for 4" SOG, the add unit price is $2.40 per sq ft and the deductive unit price is $1.80 per sq ft. For 6" SOG, the add unit price is $3.61 per sq ft and the deductive unit price is $2.70 per sq ft.

The above procedure for developing the add unit price is very slow and cumbersome, especially if many unit prices are required. An easier procedure to establish add unit prices is to develop a multiplier to be applied to the subcontractor or contractor unit prices. The formula is listed below.

Multiplier = 1 x (1 + General Condition %) x

(1 + Overhead & Profit %) x
(1 + Liability Insurance %) x
(1 + Builder's Risk %) x
(1 + Payment & Performance Bond %)

For the example shown above the multiplier = 1 x 1.05 x 1.10 x 1.02 x 1.01 x 1.01 = 1.202

Applying this factor to the concrete subcontractor's unit prices for 4" SOG and 6" SOG yields $2.40 per sq ft and $3.61 per sq ft, respectively.

As stated above, the best approach for deductive unit prices is to use the actual cost as presented or estimated, with no mark-ups applied. The reason for this approach is so the general contractor does not put himself in the position of giving up base contract mark-ups for change order work. See the example shown below.

EXAMPLE 5.7 Application of Additive and Deductive Unit Prices

Suppose a general contractor has a contract for $1,000,000 with overhead and profit of $50,000. The building is a 50,000 sq ft warehouse. The owner asked for unit prices for adding or deleting 4" slab-on-grade concrete and 6" slab-on-grade concrete. The contractor supplied the following unit prices:

- 4" Slab-on-grade   - $2.40 per sq ft
- 6" Slab-on-grade   - $3.61 per sq ft

After the contract was signed, the owner decided to change the floor of the warehouse from 6" slab-on-grade to 4" slab-on-grade. Applying the quantity of floor to the appropriate unit price results in the following:

- Deduct   50,000 sq ft of 6" SOG @ $3.61 =   ($180,500)
- Add   50,000 sq ft of 4" SOG @ $2.40 =   120,000
            Total deductive change order =   ($ 60,500)

In this transaction the contractor is giving up $2,500 in general conditions (50,000 sq ft at 5¢) and $5,500 in overhead and profit (50,000 sq ft at 11¢). The contractor's overall profit is reduced to $42,000. This is a reduction of effective overhead and profit margin from 5% (50,000 ÷ $1,000,000) to 4.47% ($42,000 ÷ $939,500).

INITIAL SCHEDULE

During the course of preparing the estimate, an obvious factor which plays an important role in the bid price is the length of time required to complete the project. The contractor's general conditions are directly affected by the project schedule. The estimator must possess a working knowledge of the project schedule should the owner dictate the time for completion and include a liquidated damages clause in the tender documents. In other instances, the owner will require the contractor to provide a schedule with the bid proposal. (Owners have also been known to dictate a liquidated damages clause to be applied to the contractor's stated completion time.)

Thus, the estimator should possess scheduling experience and skills.

The schedule that the estimator prepares (the initial schedule) should be completed after the quantity survey and associated pricing have been performed and before the contractor's general conditions are estimated.

The initial schedule should be prepared with sufficient detail to allow the estimator to accurately project the completion time. If the contractor is awarded the contract for the project, the initial schedule should be the foundation for the official project schedule.

For this to become a reality, the estimator must note all assumptions made in analyzing the initial schedule. In preparing the initial schedule, the estimator should use the following method:

- List the work as assemblies (i.e., underground building utilities, foundations, exterior walls, slab work, roofing, underground site utilities, etc.).
- Complete a schedule activity worksheet (see Fig. 5.2) for each assembly. Due to the fact that the estimator will not possess quantities for each and every activity, experience should be used to complete the schedule activity worksheet. For example, in determining the electrical work activity schedules, the estimator will have to use his best judgment in determining the time frame for each portion of electrical work since the estimator will very rarely perform a detailed electrical estimate from which the production variables are normally developed.
- Collate the schedule activity worksheets into the approximate order of the sequence in which the work is planned to be accomplished.
- Use the time-scaled activity on arrow method to develop the schedule. The nodes and arrows will represent the full duration of the activity on the time line.

The estimator must determine the level of detail to be used in preparation of the initial schedule. He should allocate enough time and effort into formulating the initial schedule to achieve a high degree of comfort with the results. Upon completion of the initial schedule, the estimator is able to evaluate the costs of schedule sensitive elements and considerations. See Example 5.8.

EXAMPLE 5.8 Project Schedule

Weeks

1 2 3 4 5 6 7 8 9 10 11 12 13 14 15 16 17 18 19 20 21 22 23 24 25 2 27

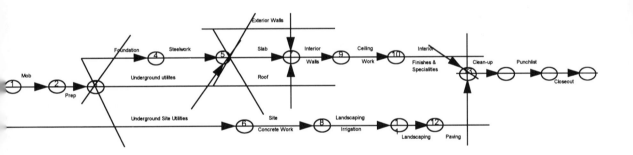

Figure 5.2 Schedule Activity Worksheet

PROJECT NAME
DATE
ASSEMBLY
ASSEMBLY QUANTITY
ASSEMBLY DURATION
OTHER ASSUMPTIONS
PRECEDING ASSEMBLY
FOLLOWING ASSEMBLY
COMMENTS

GENERAL CONDITIONS COSTS

On each project that a contractor undertakes he will incur costs that are inherent in performing the work. These costs include but are not limited to project supervision, project engineering, job site security, job site supplies and consumables, daily and final clean-up, temporary heat and power, temporary lighting, special equipment such as tower cranes and personnel hoists, safety considerations, fuel and maintenance, third-party rentals, permits and fees, quality control and testing, etc. These items collectively are known as the project general conditions. The most important aspect in determining the cost for general conditions is to estimate the duration of time that is required for the contractor to complete the project. A list of general condition items common to most projects and the manner in which their costs are calculated is listed below:

- **Project Supervision**

    This includes the project manager, superintendent, and assistant superintendents that the contractor will need to coordinate the construction of the project. Costs are calculated by taking the number of months or weeks required to build a project plus the number of weeks or months for close-out and multiplying that number by the monthly or weekly salary of the supervisor.

EXAMPLE 5.9 Cost for Project Manager and Superintendent

The project duration is 22 weeks. The close-out period is estimated at 3 weeks. The project manager earns $1,000 per week. The project superintendent earns $900 per week. The payroll burden factor is 32%. Both are full-time to the project. The estimated costs for the project superintendent and project manager are shown below:

Project manager (25 weeks x $1,000/wk x 1.32) $33,000
Project superintendent (25 weeks x $900 per week x 1.32)
.... $25,700

As a note, if the project manager had two other projects assignments, the estimate for the project manager would be $11,000.

- **Project Engineering**

  This includes the field engineer, the office engineer, draftsmen and project engineers that are bodily located on the job site. To calculate the cost, the estimator takes the number of months or weeks he estimates a person is needed on the job site and multiplies the duration by the person's monthly or weekly salary, respectively.

- **Clerical Help**

  This includes secretaries, clerks, receptionists, and timekeepers. To calculate the cost the estimator determines the number of months or weeks a person is needed on the job site and multiplies that by the individual's monthly or weekly salary, respectively.

- **Project Security**

  This item may include guards, watchmen, guard dogs and temporary fencing around the job site. To calculate costs for the guard service the estimator must determine the number of hours the guard is needed on the job site and multiply that by an hourly rate obtained from a guard service. To determine the costs for a temporary fence around the job site, the estimator should contact a fence rental service and provide them with the quantity, type of fence, and the number and size of gates required. With that information, the fence company can provide a rental price for the time the fence is required for the job site.

- **Employee Moving Expenses**

  If the project is in a location that key employees must be moved to the area of the job site, the estimator must take into consideration the cost of moving those employees.

- **Home Office Travel to Job Site**

  Once again if the project is located in an area away from the home office the estimator must make allowances for trips by upper management to and from the job site.

- **Employee Subsistence**

  If the project is of short duration away from the home office, upper management may choose to move key employees to the job site. In this case employees are usually compensated for the hardships of living away from their families. This cost may range from $100.00 to $250.00 per week per employee.

- **Job Site Office Supplies and Consumables**

    These items include costs for everyday office supplies, gloves, boots, raincoats, hard-hats, screwdrivers, hammers, etc. Costs for these items should be determined by researching historical company records.

- **Temporary Job Site Toilets**

    Costs for this item are calculated by multiplying the number of temporary toilets needed by the number of months they will be required on the job site and then that number is multiplied by the monthly rental rate. Rental rates are obtained from services handling temporary toilets.

- **Temporary Power**

    This is power required by the contractor and the subcontractors during the construction phase of the project. The electrical subcontractor for the project should include in his price all the necessary distribution and panel boards for the temporary power. However, the general contractor pays for the consumption of power. In calculating the cost for this item, the estimator should take the history of past projects into account. This item is extremely hard to quantify, and many estimators find they do not include enough money to cover the cost of temporary power. Unfortunately the discovery is not until after the project has been completed and final costs are known.

- **Temporary Heat**

    If the concrete work and interior finishes of a project are going to be performed during the winter months, the estimator must include some cost allowance for temporary heat. This is normally in the form of propane gas and propane tanks. This item is also extremely hard to quantify, and past records should be researched to identify the cost for the project. However, there are agencies that will figure the cost of these items on a rental basis.

- **Temporary Lighting**

    This item includes the cost for portable lights, tower lights, and string lights needed to provide illumination of areas that are not covered by natural lights. Once again the record of past projects should be researched in developing the cost for the project being estimated.

- **Plans and Specifications**

  The owner of the project or the architect will normally provide the contractor with a predetermined number of plans and specification. If the contractor requires more than what the owner has allowed for, the contractor must pay the reproduction costs. Normally the number of sets that is stipulated in the specifications is not enough to cover the needs of the contractor. Thus the estimator must include costs for at least 10 to 20 additional sets of plans.

- **Job Site Safety**

  This item includes the cost of all labor, material, and equipment needed to maintain a safe work site and to comply with federal and state safety regulations. This is an area that is generally grossly undervalued. This item includes activities such as those listed below:
  - Protection of openings
  - Temporary stairs
  - Temporary railings
  - Safety supplies such as hard-hats, fire extinguishers, etc.
  - Barricades

- **Job Site Clean-up**

  This activity includes the cost of all labor, material and equipment to perform daily and final clean-up. The general contractor's value for this item is to augment the daily clean-up furnished by subcontractors. Final clean-up is generally subcontracted to a janitorial service and includes mopping, dusting, waxing, window cleaning, polishing, etc. These items are also usually undervalued in the project estimate. The cost for dump fees and dumpster rental should also be included in this activity.

- **Survey and Layout**

  This is the cost to provide funds to allow a surveying company to establish lines and grades for subcontractors to locate their work. Additionally, the estimates should include material (string, stakes, batter boards, etc.) and equipment (transit, levels, etc.) costs for the project superintendent to perform layout services.

- **Protection of Existing Conditions**

    This is the cost to provide all labor, material, and equipment required for the protection of existing conditions such as trees, storm drain systems, walls, etc.

- **Protection of New Construction**

    This is the cost to provide all labor, material, and equipment necessary for the protection of new construction items such as elevator cabs, flooring, cabinet tops, walls, weather protection, etc.

- **Project Vehicles**

    This is the cost that the general contractor will incur for providing vehicles for the project. This may include the following vehicles:
    - Project superintendent's truck
    - Project manager's truck
    - General use pick-up truck
    - General use flatbed truck

    The costs calculated are based upon the number of weeks each vehicle will be used for the project and the cost to the company for operating the vehicle (excluding fuel).

- **Fuel**

    This is the cost that the general contractor will incur for providing fuel and oil for the project. The cost is calculated by estimating the number of gallons and quarts needed and multiplying that number by the appropriate unit cost. Fuel and oil are typically needed for the following pieces of equipment:
    - Job site pick-up trucks
    - Generators
    - Rented equipment

- **Personnel and Material Hoists**

    This cost is to provide personnel and material hoist(s) for buildings above two stories. The cost must include the following items:
    - Hoist rental
    - Hoist installation
    - Hoist dismantling
    - Power for hoist

- Patching holes caused by hoist bracing
- Load/Unload platforms
- Hoist foundation

- **General Purpose Labor**

    This is the cost to provide a general purpose laborer for the job site. This cost is calculated by estimating the number of hours required for a general purpose laborer and multiplying that by the hourly rate.

- **Job Site Office Equipment and Furniture**

    This is the cost to provide the project site with pertinent office equipment and furniture. This includes the following:

    - Facsimile machine
    - Video recorder
    - Copy machine
    - Camera
    - Computer
    - Tape recorder
    - Printer
    - Coffee maker
    - Desk
    - Chair
    - Plan table
    - General use tables
    - Plan holders
    - Vacuum cleaner

    If the contractor already owns this equipment and furniture, the contractor has the ability to include no costs or only a percent of the purchase cost for the items owned.

- **Petty Cash**

    This is the cost to provide the job site with petty cash needed to make direct purchases. This is intended to cover miscellaneous expenses such as stamps, coffee, supplies, meeting refreshments, cleaning supplies, office supplies, etc. This cost is usually based upon historical data.

- **Job Signs**

    This is the cost to provide and install the following signs:
    - Main project sign
    - Bulletin boards
    - Directional signs

    This cost is usually based upon historical data.

- **Project Documentation**

    This is the cost to purchase camera film, video film, audio tapes, and to develop the film. This cost is also usually based upon historical data.

- **Temporary Roads and Parking**

    This is the cost to grade and install gravel for temporary roads and parking. The cost is generally based upon the square yards of temporary roads and parking that are required.

- **Permits and Fees**

    This is the cost to pay local government agency building permit fees. This cost is calculated from data provided by the local government agency.

- **Third-Party Rentals**

    This is the cost to provide rentals for equipment needed by the general contractor but either not owned or not available. This includes the following:
    - Forklifts
    - Cranes
    - Generators
    - Trucks

    The costs are developed by determining the days, weeks or months the equipment is needed and multiplying that by the appropriate unit cost. Additionally, the routine maintenance cost of the rented equipment must be included in the bid.

INSURANCE CONSIDERATIONS

Every project which the estimator looks at will require some form of insurance. It is important for the estimator to have a working knowledge of the types and costs of insurance that the owner or architect specifies as a contractual requirement.

Public Liability and Property Damage Insurance

Public liability insurance covers the liability of the insured for negligent acts resulting in bodily injury, disease, or death to others (other than employees of the insured).

Property damage insurance covers the injury to or destruction of tangible property, including loss of use resulting therefrom, but usually not including property which is in the care, custody, or control of the insured.

The premium for this insurance coverage is calculated on a percentage of the base wages of the workmen, amount of subcontract work, and home office payroll.

Prior to the start of each year of insurance coverage, the contractor's insurance company requests an estimate for the following costs:

- Labor to be performed by the contractor during the year.
- Amount of work to be subcontracted during the year.
- Labor used to manage the projects.
- Office overhead labor for the year.

Insurance premium factors are applied to each of the estimated costs listed above to develop an estimate for the contractor's liability insurance premium for the coming year.

One method for determining the amount for liability insurance to apply to each project and change orders is described below.

Assume that ABC General Contractor has provided the insurance company with the information requested and subsequently been given the following premium schedule:

EXAMPLE 5.10 Liability Insurance Rate Calculation

Activity	Estimated Expense	Factor	Amount
Direct Labor	$1,000,000	6.0%	$ 60,000
Subcontractors	6,000,000	0.5%	30,000
Project Management	490,000	1.0%	4,900
Office Overhead	300,000	0.5%	1,500
Total Estimated Premium			$96,400

In order to develop a factor to apply to each project, the contractor must estimate the amount of contracts to be executed within the coming year. This can be accomplished by adding the labor, subcontract, and figures shown above to the following.

- **Equipment rentals**

- Estimated material costs
- Payroll burden costs
- Overhead and profit
- Builder's risk insurance premiums
- Project general conditions

To ensure that the entire premium is covered during the year, the estimate should be on the conservative side. At the end of each coverage year, the insurance company will audit the contractor's books to determine actual costs of the categories that apply to premium factors. During the year, the contractor makes monthly installment payments against the estimated premium. At the completion of the audit, the contractor will either owe more or have a credit.

ABC General Contractor's volume estimate is shown below.

Item	Estimated Amount
1. Direct Labor	$1,000,000
2. Subcontractors	6,000,000
3. Equipment Rentals	200,000
4. Material Costs	1,000,000
5. Payroll Burden (40% of 1)	400,000
6. Subtotal of Direct Project Costs	8,600,000
7. Project General Conditions (6% of 6)	516,000
8. Subtotal	9,116,000
9. Overhead & Profit (6% of 8)	547,000
10. Subtotal	9,663,000
11. Builder's Risk Insurance ($.25/100 of 9)	25,000
Total of Estimated Volume less Liability Insurance	$9,688,000

The figure of item 11 is divided into the total estimated premium to determine the liability insurance factor. For ABC General the factor is 1% ($96,400 ÷ $9,688,000).

Some contractors add other insurance items such as equipment insurance, product liability insurance, home office insurance, etc., to the liability insurance premium to establish an overall insurance factor to apply to each bid to capture all insurance costs as project costs.

Products Liability

Provides insurance for liability imposed for damages caused by an occurrence arising out of goods or products manufactured, sold, handled, or distributed by the insured or others trading under his name. Occurrence must happen after product has been relinquished to others and away from premises of insured. Rates are provided by the contractor's insurance agent.

Owner's Protective Liability

Provides insurance to cover the liability of the owner for injuries to the public arising out of operations of independent contractors. Rates are based upon the total estimated cost of the project and are obtained from the contractor's insurance agent.

Independent Contractor's Liability

Provides liability coverage for claims caused by occurrences and based upon a contractor's contingent liability usually resulting from the operations of his subcontractors. Rates may be obtained from the contractor's insurance agent.

Automobile Liability Insurance

Provides insurance for bodily injury and property damage for owned vehicles, non-owned vehicles, hired vehicles, or independent contractors. Rates may be obtained from the contractor's insurance agent.

Builder's Risk Insurance

Provides protection for the structure and appurtenances being built during the course of construction. Builder's risk premium rates are based on the total estimated value of the project. This is normally written on a standard fire policy with extended coverage for windstorm, hail, explosion, riot, smoke damage, vandalism, theft, and malicious mischief. Rates may be obtained from the contractor's insurance agent. The insurance agent will require the following information: type of construction, estimated cost of construction, construction duration, location of project, street address of project and owner's name. The rate is stated as dollars per one hundred dollars of contract value per year. For example, if the estimated cost and completion time for a project were $5,000,000 and two years, respectively, and the rate provided by the insurance agent is ten cents/one hundred dollars, the builder's risk premium would be calculated as follows:

EXAMPLE 5.11 Builder's Risk Premium Calculation

Builder's Risk Premium = $.10/year x $5,000,000.00/$100.00 x 2 years = $10,000.00

In some cases, the owner will provide builder's risk insurance for the project. In that situation the estimator must determine if there is any residual risk to the contractor (i.e., a deductible which is the responsibility of the contractor; exclusions such as earthquake, floods, acts of war, and so forth; and incorrect amount or duration of coverage) and determine the cost to mitigate those risks. The mitigation costs must be included in the bid price.

TESTING COSTS

Testing costs are defined as those costs required to test and inspect various components of a project. Testing and inspection includes, but is not limited to, the following activities.

- Testing earthwork for compaction, moisture content, elasticity characteristics, etc.
- Testing concrete for slump, air content ,and compressive strength.
- Testing potable water systems for toxicity.
- Visually inspecting welds.
- Testing bolted connections for appropriate torque.
- Visually inspecting a roofing subsystem.
- Testing grout for compressive strength.
- Pressure-testing piping systems.
- Performing special inspections of building systems, i.e., structural, electrical, mechanical, plumbing, roofing, etc.
- Testing asphalt for compressive strength.

The tests required for each project as well as the party responsible for paying for the tests are found in the conditions (general, special, etc.) and the specifications of the project documents.

In determining the total costs to be included in the bid, the estimator should quantify the tests needed and request a budget number from appropriate testing agencies.

Items to include in the budget for testing are listed below:
- Field technician's time
- Laboratory test costs
- Travel time
- Cost to produce reports
- Reimbursables associated with testing
- Equipment costs

In order to avoid a conflict of interest and reduce risk, the contractor should contract with independent testing agencies to perform all project-related testing.

Chapter 6

BID DAY PROCEDURES

PREPARING FOR BID DAY

Preparing for bid day is much like studying for a test. The estimator who has not performed his homework will more than likely score poorly against the competition. On the other hand, the estimator who knows the job will probably fare better against the other bidders.

In the days prior to bid day the estimator should perform the items described below:

- Review specifications at least twice.
- Thoroughly review each page of the drawings to ensure each item has been either quantified and priced or will be covered by a subcontractor or material supplier.
- Make sure that an ample number of suppliers, vendors, or subcontractors have been contacted and will be quoting items other than the labor activities of work that will be accomplished by in-house forces.
- Prepare a scope of work checklist for each major activity that will be subcontracted. Provide each person who will be receiving quotations with the lists. See the example below.

    Section 7100 - Waterproofing
    - Below-grade waterproofing
    - Caulking and sealing of doors
    - Under slab waterproofing
    - Planter wall and base waterproofing

- Prepare an itemized list of materials to be included in specific supplier quotations. See the example below:

    Section 5100 - Structural Steel Suppliers
    - Structural steel

- Joists
- Metal roof deck
- Metal floor deck
- Embedded steel
- Grating
- Steel stairs
- Ladders
- Anchor bolts
- Steel lintels
- Sag rods
- Girders
* Prepare the bid documents and obtain required signatures.
* Obtain the bid bond from the surety company if one is required.
* Prepare submittals to be provided. This may include the following items:
  - Project schedule
  - Project payment request schedule
  - Manpower schedule
  - Resumes of proposed project personnel.
  - Proposed work plan
  - Conformance documents required by government agencies.

ADDENDA

Whenever the project designer finds that it is desirable, necessary, or advisable to make a change in the tender documents after initial issuance, he will issue an addendum.

Since addenda are changes to the drawings or specifications, the estimator must determine the effect that the addenda has on the bid price. He must also be sure that any subcontractor or supplier affected by the addenda receives and acknowledges having seen the addenda.

It is a good rule to show the revisions made by the addendum on the set of plans and specifications that the estimator is or will be using in the quantity take-off phase. This is easily accomplished by making a copy of the addendum and cutting and pasting (or stapling) the portions of the addendum to the specification or drawing affected.

The bid forms will have a section for the contractor to write in the addenda he has received for the project. Failure to acknowledge the addenda may result in the bid being declared non-responsive and disqualified. Thus, it is suggested that the estimator place a call to the designer one or two days prior to the bid date to check that all addenda have been received and that no others will be issued.

BID BUCKET

The estimator should set up a bid bucket for the project (see Fig. 6.1). The bid bucket consists of a legal-size box with hanging folders for each trade for which the estimator expects to receive bids. Hanging folders should also be inserted into the bid bucket for sections including abstracts, addenda, estimate, bid forms, and specifications.

The purpose of the bid bucket is to categorize the bids received into similar scopes of work so the estimator running the spreadsheet will have quick access to compare pricing and scopes of work between subcontractors and suppliers.

Placing the addenda, abstracts, estimate, bid forms, and specifications in the bid bucket allows the estimator quick and easy access to that information.

Figure 6.1 The Bid Bucket

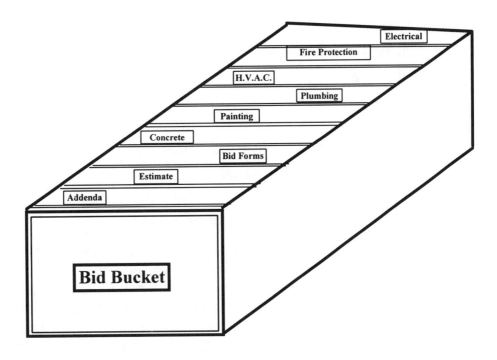

THE ESTIMATE SPREADSHEET

One of the most important tools used by the estimator to develop the final costs of a project is the estimate spreadsheet (see Fig. 6.2).

A project is composed of independent components. The estimator has the responsibility to ensure that each and every component is identified and that each and every component will be covered by estimates performed by the estimator or by bids received from subcontractors and/or suppliers.

The estimator must use the research from the initial preparation phase to establish the components of the project which will be placed on the spreadsheet. Each component for which independent bids are anticipated should be listed separately. An example of this would be listing "Door Labor" (labor to hang doors) separate from "Wood Doors" and "Hollow Metal Work." These products are typically bid by three independent businesses.

The estimate main summary is developed using a spreadsheet with the following columns:

- Specification number (if applicable).
- Description of work activity.
- Labor costs.
- Material costs.
- Equipment costs.
- Subcontractor costs.
- Other Expense
- Total costs vendor/sub name.
- Adjustment columns which will include the following subheadings.
    - New vendor/sub name
    - New price
    - Add or deduct from previous price

On the day prior to bid day, the estimator should transfer all the pricing from the estimate worksheets to the main summary spreadsheet (if the bid proposal contains alternates, a separate spreadsheet should be established for each alternate). For the components of the project that a detailed estimate was not performed, a plug number should be inserted in the appropriate column. Under the sub/vendor column each bidding entity should be identified. This should be the name of the vendor, subcontractor, general contractor (GC), or plug if a firm price has not been identified. In some cases where the general contractor is performing the work and buying the material, both GC and the vendor's name should be listed.

Inserting a number in all applicable spaces will allow the estimator to total the costs down to adjustment number one prior to bid day.

On bid day, as the proposals are received, the estimator must compare the new bids with the amounts shown on the spreadsheet for the same scope of work. (See the section on evaluating trade bids.)

As better or more defined bids are received, the adjustment columns are completed. The adjustment columns provide spaces for the estimator to insert the new subcontractor or vendor's name, the new price, and the change (+ or -) from the price indicated on the spreadsheet.

An estimate spreadsheet will normally contain two to five adjustment columns. The estimator should predetermine the time that each adjustment column will be added up and then enter an adjustment onto the main total column.

For example, the estimator might determine that the adjustment columns should be completed at 11:00 a.m., 12:00 p.m., 1:00 p.m., and 1:30 p.m. for a 2:00 p.m. bid time.

After the last adjustment column is completed, the estimator should have a feeling for what the final bid should be and where the possible risks of the project are located. The overhead and profit, insurance, and bonds can be added at this point. New bids should be analyzed and adjusted on independent lines with careful notes beside each line identifying the new costs, old costs, and new subcontractor or supplier. At some point the final bid number(s) must be determined and given to the person who will be turning in the bid proposal.

Figure 6.2 Estimate Spreadsheet

ESTIMATE SPREADSHEET
PROJECT: Real Estate Office Tenant Improvement Bid Time - 2:00 P.M. Bid Date - 12/28/94

SPEC NO	Activity	Labor	Materials incl sales tx	Equipment incl sales tx	SUBS	Other-incl sales tx	Totals	Vendor	Adjustment No. 1 New Price	Add/Cut	New Vendor	Adjustment No. 2 New Price	Add/Cut	New Vendor
1000	General Conditions	2000	1000		1000	1500	5500	G.C.						
6100	Rough Carpentry	1000	600				1600	G.C.						
6400	Cabinets				6000		6000	R.J.						
7200	Insulation				4000		4000	PLUG	6250	2250	EXCEL	6114	-136	ABC
8100	H.M. Frames	500	1500			100	2100	GC/PLUG	1475	-25	VALUE			
8200	Wood Doors	1500	3000			200	4700	GC/PLUG	3260	260	VALUE			
8700	Door Hardware	1000	2000			200	3200	GC/PLUG	2170	170	VALUE			
8800	Glass				2000		2000	PLUG	1960	-40	REFLECT			
9200	Drywall				11000		11000	PLUG	14520	3520	EXCEL	13275	-1245	ANDYS
9300	Ceramic Tile				3000		3000	PLUG	2760	-240	BAY			
9500	Acoustical Ceiling				7000		7000	PLUG	6420	-580	JONES	6110	-310	JOHNS
9600	Soft Flooring				8000		8000	PLUG	8120	120	SMITH			
9900	Painting				7000		7000	PLUG	6380	-620	BRADS			
15400	Plumbing				6000		6000	PLUG	7200	1200	JIMS	6690	-510	SIDS
15500	Fire Protection				5000		5000	PLUG	5435	435	HARBOR			
15800	H.V.A.C.				8000		8000	PLUG	7640	-360	TURNER			
16000	Electrical				11000		11000	PLUG	13400	2400	ARC	12345	-1055	SWIFT
	Subtotals	6000	8100	0	79000	2000	95100			8490			-3256	
	Adj. No. 1						8490							
	Adj. No. 2						-3256							
	Adj. No. 3	HVAC - RAY'S AC BID $7110 CUT $530					-530							
	Adj. No. 4	CERAMIC TILE - PORT'S BID $2570 CUT $190					-190							
	Subtotal						99614							
	Profit	12.50%					12452							

RECEIVING AND EVALUATING SUBCONTRACTOR AND SUPPLIER BIDS

The solicitation of subcontractor and supplier bids, as discussed elsewhere, is completed upon the receipt of the proposer's bids on or before bid day.

Receiving and analyzing sub-bids properly is an important cog in putting together a competitive and profitable estimate.

Sub-bids will normally be received in one of the following two fashions:
- A written proposal from the subcontractor in person, through the mail or by facsimile machine.
- A verbal bid received over the telephone.

With today's technological advances, the contractor should have at least one facsimile machine available to receive bids.

Since verbal communication is not reliable and one person's interpretation is usually different from another person's, it is very important not to rush through the taking of a bid over the telephone.

The person taking telephone bids should have an ample supply of telephone bid forms handy at all times on bid day (see Fig. 6.3). When receiving a telephone quotation the following information should be taken:
- Name of project subcontractor/supplier is bidding on (since they bid several projects each week, it is not uncommon to get projects mixed-up).
- Name of company (when bidding out of town or receiving a bid from a new company the estimator should also write the address of the company on the bid form).
- Name of person giving bid.
- Telephone number of the company.
- The number of addenda acknowledged by the firm.
- Specific exclusions of the subcontractor.
- The next step involves taking down the subcontractor's scope of work, inclusions, exclusions, base bid price, and alternate pricing. It is extremely important to write down all the information given by the sub so the chief estimator has all the facts in order to properly analyze the sub's bid relative to others who are bidding on the same categories.

After all the information has been received, the person taking the quote should confirm all the information received is correct by reading the bid back to the subcontractor/supplier.

Until one hour before bid time a person taking subcontractor/supplier bids should accumulate several quotes before turning them over to the chief estimator. During the last hour, each bid should be given to the chief estimator as it is received.

Figure 6.3

SUPPLIER/SUBCONTRACTOR TELEPHONE QUOTATION FORM					
DATE					
NAME OF PROJECT					
NAME OF BIDDER					
ADDRESS OF BIDDER					
BIDDER'S TELEPHONE NO.					
ADDENDA ACKNOWLEDGED					
BASE BID AMOUNT					
ALT NO. 1	ALT NO. 2	ALT NO. 3		ALT NO. 4	ALT NO. 5
ALT NO. 6	ALT NO. 7	ALT NO. 8		ALT NO. 9	ALT NO. 10
SCOPE OF WORK					
EXCLUSIONS					

SCOPE OF WORK CHECKLISTS

The purpose of this section is to provide a checklist of questions to ask subcontractors and suppliers to verify the scope of work included in the subcontractor's or supplier's bid proposals.

- **General Questions for all Subcontractors or Suppliers**
    - Is work bid in accordance with plans, specifications, general conditions, special conditions, addenda, etc.?
    - How many addenda has the subcontractor/supplier reviewed?
    - Is the subcontractor/supplier aware of the liquidated damages clause (if applicable)?
    - Does the subcontractor/supplier know of any long lead time, materials, or equipment which may affect the required completion time for the project?
    - Is the subcontractor/supplier aware of any materials specified which have been discontinued?
    - What specification sections is the subcontractor/supplier specifically bidding?
    - What are the subcontractor's/supplier's exclusions?
    - What is the subcontractor's cost to bond the project?

- **Site Grading Subcontractor**
    - How many move-ins?
    - Cubic yards of excavation?
    - Cubic yards of backfill?
    - Is clearing and grubbing included?
    - Is topsoil included?
    - Is fine grading of pavements included?
    - Is fine grading of slabs included?
    - Is over excavation for foundations included?
    - Has the subcontractor reviewed the soils report?
    - Is foundation excavation included?
    - Is playground sand included?
    - Is underslab base course included?
    - Is base course for paving included?
    - Is rock excavation included?
    - Is excavation for site utilities included?
    - Are haul fees included?
    - Is survey and layout included?

- **Asphalt Paving Subcontractor**
    - Is fine grading included?
    - Are pavement markings included?
    - Are parking lot signs included?
    - Is soil treatment included?

- **Site Utilities Subcontractor**
    - Are concrete thrust blocks included?
    - Are tap fees included?
    - Is site sewer system included to within 5 ft of building(s)?
    - Is water service included to within 5 ft of building(s)?
    - Is storm drainage included?
    - Are cast-in-place catch basins included?
    - Is underground fire protection included to flange(s) inside building(s)?
    - Is site natural gas piping included?
    - Has the subcontractor reviewed the soils report?
    - Where is excess excavation being taken?
    - Are water meter fees included?
    - Is water meter installation included?
    - Is landscape irrigation system included?
    - Is chlorinization of water system included?

- **Landscaping Subcontractor**
    - Is playground sand included?
    - Is water meter installation for irrigation included?
    - Are water meter fees included?
    - Is topsoil included?
    - Is rock work included?
    - Are brick pavers included?
    - Are brick mowing strips included?
    - Are concrete mowing strips included?
    - Is backflow prevention included?
    - Is soil treatment included?
    - Is site furnishing(s) included?
    - Are tree gratings included?

- **Site Concrete Subcontractor**
    - Are concrete driveways included?
    - Are all sidewalks included?
    - Are all concrete curbs included?
    - Are all valley gutters included?
    - Are light pole bases included?
    - Is reinforcing steel for concrete work included?
    - Are cast-in-place catch basins included?
    - Are cast-in-place manholes included?
    - Are concrete mow strips included?
    - Is excavation and backfill included for vertical curbs?
    - Are steel pipe billiards included?
    - Are parking lot signs included?
    - Are footings for site walls included?
    - Are exterior equipment pads included?

- **Concrete Subcontractor**
    - What site work concrete is included?
    - Is reinforcing steel included?
    - Is fine grading for slabs included?
    - Is perimeter insulation included?
    - Is vapor barrier under slab included?
    - Are concrete equipment pads included?
    - Are joint sealers included?
    - Is caulking of tilt-up panels included?
    - Are shop drawings for tilt-up panels included?
    - Is engineering for tilt-up panel lifting hardware included?

- **Masonry Subcontractor**
    - Is flashing of masonry included?
    - Are precast sills, etc., included?
    - Is integral waterproofing included?
    - Are brick pavers included?
    - Are brick mow strips included?
    - Is reinforcing steel included?
    - Are trash enclosures included?
    - Are site masonry fences included?
    - Is shoring of masonry openings and beams included?
    - Is installation of embedded metals included?
    - Is installation of sill metal angles included?

- Is setting of hollow metal door and window frames in masonry walls included?
- Is caulking of expansion joints included?
- Are footings included for masonry fences?

- **Structural Steel Subcontractors/Suppliers**
  - Is installation of steel included?
  - How many tons of structural steel are included?
  - How many squares of metal deck included?
  - Are ladders included?
  - Is steel railing included?
  - Is aluminum railing included?
  - Are shop drawings included?
  - Are expansion joints and expansion joint covers included?
  - Are foundry items included?
  - What items are f.o.b. job site only?
  - Are wood to wood steel connections included?
  - Are anchor bolts included?
  - Are roof hatches included?
  - Are steel joists included?
  - How many tons of steel joists are included?
  - Is structural studwork included?

- **Rough Carpentry Subcontractors**
  - Is all lumber included?
  - Are wood-to-wood steel connections included?
  - Are masonry-to-wood steel connections included?
  - Are concrete-to-wood steel connections included?
  - Are wood trusses included?
  - Are wood joists included?
  - Is wood blocking included?
  - Is wood siding included?
  - Is steel venting included?
  - Is all wood framing included?

- **Finish Carpentry Subcontractor**
  - Are wood door frames included?
  - Are wood doors included?
  - Are cabinets included?
  - Are countertops included?
  - Is cultured marbled work included?
  - Is finishing of trim pieces included?

- Is wood base included?
- Is installation of all doors and door hardware included?

- **Insulation Subcontractor**
  - Is perimeter insulation included?
  - Is insulation for sandwich concrete slabs at freezers included?
  - Is sound insulation included?
  - Is roofing insulation included?
  - Is vapor barrier under slab included?
  - Is insulation under base plate included?
  - Is sound-deadening board included?

- **Waterproofing Subcontractor**
  - Is exterior watersealing included?
  - Is caulking included?
  - Is joint sealing included at concrete joints?
  - Is concrete sealer included?
  - Is all below ground waterproofing included?
  - Is flashing included?
  - Is caulking of expansion joints included?

- **Roofing Subcontractor**
  - Are roof crickets included?
  - Is insulation included?
  - Are roof jacks included?
  - Are pitch pockets included?
  - Is steel roofing included?
  - Are asphalt shingles included?
  - Are wood shingles included?
  - Is testing included?
  - What type of warranty is included?
  - Are walking pads included?
  - Are flashing and general sheet metal included?

- **Wood Door Supplier**
  - How many doors are included?
  - Are the doors pre-machined?
  - Are the doors prefinished?

- **Door Hardware Supplier**
    - Is hardware for entrance doors included?
    - Is hardware for specialty doors included?

- **Glass And Glazing Subcontractor**
    - Is aluminum framing included?
    - Is hardware for entrance doors included?
    - Is cleaning of glass included?
    - Are mirrors included?
    - Is caulking associated with glass, glazing, and aluminum framing included?
    - Are lite kits for hollow metal doors included?

- **Stucco Subcontractor**
    - Is metal framing supporting stucco included?
    - Is drywall backing stucco included?
    - Is vapor barrier included?
    - Is scaffolding included?

- **Drywall Subcontractor**
    - Is structural studwork included?
    - Is wonderboard behind ceramic tile included?
    - Is sound-deadening board included?
    - Is scaffolding included?
    - Is spray on textured ceiling included?
    - Is patching of existing walls included?
    - Are tape and texture included?
    - Is wood blocking or metal blocking included?
    - Is setting of hollow metal door and window frames included?

- **Ceramic Tile Subcontractor**
    - Is wonderboard behind ceramic tile included?
    - Are marble or cultured marble thresholds included?

- **Acoustical Ceiling Tile Subcontractor**
    - Are corners mitered or butted?
    - Is scaffolding included?
    - Are cutouts for sprinkler heads included?

- **Soft Flooring Subcontractor**
    - Are mopping and waxing resilient flooring included?
    - Is vacuuming of carpet included?
    - Is removal of existing carpet included?
    - Is sweeping existing surface included?
    - Is filling holes in the existing floor surface included?
    - Is preparing the existing surface for new flooring included?
    - Is wood base included?

- **Painting Subcontractor**
    - Is wall covering included?
    - Is scaffolding included?
    - Is removal of existing wall covering included?
    - Is sizing of drywall to receive wall covering included?
    - Is touch-up included?

- **Elevator Subcontractor**
    - Are sill angles included?
    - Is flooring in the cab included?
    - What is the lead time after approval of shop drawings?

- **Swimming Pool Subcontractor**
    - Is the pool deck included?
    - Where does the swimming pool subcontractor pick up water supply and electrical service?
    - Is rock excavation included?

- **Plumbing And Piping Subcontractor**
    - Is the plumbing permit included?
    - Are condensate drains included?
    - Are valves included for HVAC piping?
    - Is plumber picking up water, sewer, and gas from 5ft outside building?
    - Is caulking of plumbing work included?
    - Are tap fees included?
    - Is piping insulation included?
    - Is pipe identification included?
    - Are coordination drawings included?
    - Are as-built drawings included?
    - Are equipment pads included?
    Is rock excavation included?

- **Fire Protection Subcontractor**
    - Is the permit included?
    - Is the design included?
    - Are shop drawings included?
    - Are sprinkler heads centered on acoustical ceiling tile?
    - Is the fire annunciator panel included?
    - Is fire protection subcontractor picking up piping from flange inside the building?
    - Are coordination drawings included?
    - Are as-built drawings included?
- **HVAC Subcontractor**
    - Are condensate drain lines included?
    - Are temperature controls included?
    - Is low-voltage wiring for temperature controls included?
    - Are shop drawings for ductwork included?
    - Are coordination drawings included?
    - Is testing and balancing system included?

    Are concrete equipment pads included?
- **Electrical Subcontractor**
    - Is connecting the general contractor's office trailer to a power source included?
    - Is a temporary power pole included?
    - Is the temporary power distribution included?
    - Is temporary lighting included?
    - Is low-voltage wiring for temperature controls included?
    - Are smoke detectors included?
    - Is fire alarm system included?
    - Is security system included?
    - Are the electrical permits included?
    - Are utility company fees included?
    - Are concrete light pole bases included?
    - Are concrete equipment pads included?

PROFIT CONSIDERATIONS

For the purpose of this discussion, profit is defined to be that amount of money which exceeds the project costs. As stated, this definition also includes the contractor's general and administrative (G & A) costs.

In order to properly evaluate the amount of profit to include in each project bid, the officers of the construction company need to perform financial planning and set goals before the beginning of each fiscal year. The following factors should be considered:

- Projected G & A (overhead) expenses (based upon past fiscal years).
- Goal for volume of work to be billed out for fiscal year.
- Goal for cash to be added to the company for future growth.
- Amount of profit on projects that are carried over from the following fiscal year.

EXAMPLE 6.1 Company Overhead Rate Calculation

Landmark Construction Company has accumulated the financial planning information shown below:

- **G & A expense for previous year was $200,000.**
- **G & A has increased 5% per year for last four years; therefore projected G & A is $210,000.**
- **The company has maintained an 8% growth rate in terms of billings for the past four years.**
- **The billings for the past fiscal year were $10,000,000.**
- **Thus the billings for the coming fiscal year are expected to be $10,800,000.**
- **The company is carrying $50,000 of unearned profits on projects into the new fiscal year.**
- **The company plans to increase net worth by $50,000 this fiscal year ($100,000 pretax dollars).**
- **The two owners of the company have set a goal of making a $50,000 bonus each for the fiscal year.**
- **The net worth of the company is $200,000.**
- **The company expects to be overbilled by $100,000 each month.**
- **The officers have set a goal to give $50,000 in bonuses to employees.**

Thus the following conclusions can be reached.
- **Percent of volume needed to cover G & A ($210,000) ÷ $10,800,000 = 1.945%.**
- **Percent of volume needed to cover increase in net worth ($100,000 ÷ $10,800,000) = 0.926%.**
- **Percent of volume needed to cover employee bonuses ($50,000 ÷ $10,800,000) = 0.463%.**
- **Percent of volume needed to cover bonuses ($100,000 ÷ $10,800,000) = 0.926%.**
- **Reduction due to carryover profits ($50,000 ÷ $10,800,000) = 0.463%.**

Thus in order to meet the company's fiscal goals, a minimum of 3.796% mark-up should be applied to the first $10,800,000 in volume that the company bills during the coming fiscal year. (Note that this percentage can be reduced by subtracting the percentage calculated from the amount of interest earned on items nine and ten above.)

Each month the goals should be adjusted according to how the company is performing financially. For example, one school of thought is to increase the mark-up percentage once the financial goals are covered; the other school is to reduce the mark-up percentage (to obtain more pure profits).

Once the minimum profit has been established, the officers of the company should review each project individually to determine the actual profit to apply to the costs of the project.

The following items should be considered in determining the profit for a given project:
- Complexity of the project.
- Duration of the project.
- Reputation and credibility of apparent low subcontractors and suppliers.
- Amount of labor the general contractor will perform himself (as a rule of thumb the minimum final mark-up should be greater than 30% of the general contractor's labor estimate).
- Time of year the project will be breaking ground.
- The anticipated risks and problems that may be encountered.
- Evaluation of "bid" market from previous estimates.
- Reputation of the architect/engineer with regards to fairness.
- Reputation of the owner with regards to timely payments.

- Contractor's current workload. (If the workload is high, the contractor might add a percent or two; if the workload is low the contractor might take-off a point or two.)
- Location of the project.
  - Reputation of the local building inspection department.
  - Access to the site.
  - Availability of good craftsmen.
- Reputation of the owner in dealing fairly with contractors.

A construction company cannot remain in business without attaining a profit on construction contracts. As such, the amount of profit applied to a bid project requires due diligence on the part of the construction company executives.

TAXES

Taxes may be assessed for almost any reason that local, state, and federal governments want to use for producing revenue. Thus, the estimator must carefully study the tax situation in the locality of the project. He must also stay abreast of changing tax situations (just because Metropolis had a 6% sales tax rate two years ago does not mean it is still 6% today).

Information concerning taxes in a specific locality should be obtained from both the Department of Revenue of the city or county and the State Treasurer's Office.

Taxes which the estimator may encounter are listed below:
- Sales tax on purchase of goods, materials, and services.
- Use tax on purchase of goods, materials, and services.
- Business tax on contract value.
- Manufacturing tax on items prefabricated off the construction site.
- Gross receipts tax on the contract value.
- Road and highway taxes.
- Tariffs and duties on the value of imported goods.
- Corporation franchise taxes on capital employed in a state.
- Fuel tax greater than normally required.
- Property tax on equipment brought into state.
- Occupational license tax.

In addition to local taxes, the contractor must also include the cost of payroll taxes in his bid. (See the section on labor costs.)

COSTS THAT SHOULD BE CONSIDERED

In determining the final bid to submit for a project, construction company owners and executives continually seek to trim the "fat" out of the numbers in order to seek out a leaner bid price with the intent of being "more competitive."

In many instances, this cannibalism of the estimate exposes the construction company to risks that are not evident on bid day.

The purpose of this section is to discuss those areas for which the contractor should not cut costs and also to introduce consideration to add costs to the bid which are traditionally excluded.

Subcontractor/Supplier Pricing (Buy-Out Cut)

In their hunger to be the low bidder many contractors arbitrarily reduce the price of individual subcontractor/supplier quotations. Others introduce a line item to the spreadsheet entitled "Buy-out Cut." In both of these situations the contractor hopes (if he is the successful low bidder) to make up these cuts by "negotiating" the prices submitted by subcontractors and suppliers on bid day, or by "shopping" subcontractors and suppliers bid prices to other subcontractors and/or suppliers (including to those who did not submit a quote on bid day).

In reality the contractor is reducing the bid price overhead and profit. This policy could also put the contractor in the situation of getting the reputation of condoning unethical practices.

General Conditions

Another bid day target for the price-cutting hatchet is the contractor's general conditions. Most of the general conditions costs are based upon time-related issues. As the fervor of bid day builds, the competitive juices begin to flow to the contractor's brain. The contractor's competitive attitude may make him feel that he can chop time off the schedule prepared by the estimator for the project since his company can outperform all of his competitors. This may lead to an arbitrary cut of the general conditions costs.

For the contractor this reduction is usually double jeopardy as most general contractors overrun the general conditions budget on every project.

This cut is also, in reality, reducing the bid price overhead and profit.

Overhead and Profit

The component for the bid that is most often reduced is the contractor's overhead and profit. Before cutting the profit margin, the contractor must consider several factors. These factors include the following:
- Is the overhead covered for the year?
- What are the risks involved in the project?
- Have other reductions (subcontractor and/or general conditions cuts) already been taken?
- Are there any holes in the estimate?

The temptation to reduce overhead and profit on bid day grows exponentially as bid time approaches. The contractor should resist this temptation and remember the considerations used in originally determining the profit margin that should be applied to the project. (See the section on profit considerations.)

The next portion of this section introduces cost components that the contractor should include in the bid, but are traditionally excluded or disregarded.

Legal Fees

During the course of the fiscal year a construction company will require the services of an attorney. These costs traditionally are not budgeted on either an overhead basis or on a project basis. Thus, the costs for legal counsel depletes the company's profits. As a result of this historical reduction of profits, contractors should consider including legal fees in bids through either increasing the overhead pool or including legal fees in individual bids.

Insurance Deductibles

Most contractors purchase liability insurance and builder's risk insurance with a clause for a deductible. Contractors should consider including the cost for insurance deductibles in their bids. This can be accomplished by either increasing their insurance pool (thus increasing the rate used for liability and builder's risk insurance) or adding this cost to individual bid projects.

Punchlist

During the close-out of a project, one of the last activities which must be accomplished before the retention is released to the contractor is compilation and completion of a punchlist.

The final punchlist can range from less than one page to many pages of items which require correction (such as stains on carpet, hand prints on acoustical ceiling tile, paint touch-ups, dented frames, etc.) or attention (such as submittal of operation and maintenance manuals to owners, submittal of lien waivers, submittal of certificate of occupancy, and submittal of as-built drawings).

This process nearly always runs beyond the original project schedule and thus requires additional supervision and labor which is typically not included in the bid price.

In order to reduce the risk for this issue, contractors should consider adding a line item to their bid spreadsheets for punchlists. Another possibility is to put an activity in the general conditions for punchlist.

Interest for Working Capital Funds

Historically, contractors have excluded the costs for interest since the requirements to borrow working capital funds have been minimal. Subcontractors and suppliers are generally paid when payment is received from the owner.

As financial institutions and owners (and their attorneys) are becoming wiser to the workings of the construction industry, they are realizing that "pay-when-paid" contract clauses between the contractor and subcontractor are a potential for high risk. The possibility exists (and is even borne out nationwide every day) that the contractor will/does use project funds for purposes other than paying subcontractors and suppliers. This results in one or more of the following.

- The owner having to pay subcontractors or material suppliers directly (thus the saying "The owner pays twice").
- Subcontractors and suppliers requesting payment from the bonding company of the contractor, if the project is bonded.
- Liens being filed on the property by the subcontractors and suppliers who haven't been paid.

One method that a few owners are using to reduce their risk in paying the general contractor and accepting conditional lien releases is to require the general contractors to pay their subcontractors and suppliers (and thus require subcontractors to do the same) prior to issuing a progress payment. Since most general contractors do not maintain the working capital to pay all their suppliers and subcontractors before receiving monthly progress payments, contractors will be forced to borrow the working capital and pay interest for a period ranging from a few days to a few weeks.

As such, contractors dealing with owners who will not accept conditional lien waivers must include interest costs in their bids.

Subcontractor/Supplier Risk Factor

During the bid process the obsession to use the lowest possible subcontractor/supplier bid price is overwhelming. The tendency in bidding a project is to rely on the lowest subcontractor/supplier bid in each category.

Some of the reasons for the obsession and tendency are listed below.

- Other contractors bidding the project will use the lowest numbers for each activity.
- The time spent putting the bid together will be wasted if the lowest bids are not used.
- There is no chance to be the successful bidder if the lowest subcontractor/supplier bids are not used.
- If the low bidder does not honor his number there is legal recourse against that subcontractor/supplier to reduce the risk of using the bid.

Although each of those "reasons" may sound perfectly logical on bid day, one of the main reasons more contractors are going out of business today than ever before is from accepting subcontractor/supplier bids that are too low.

Many general contractors endorse and use the practice of calling subcontractors on bid day who are too low with hopes that the subcontractors will either pull their bids or at least raise the number (which gets into an ethical situation if the subcontractor/supplier purposely submits a low number with hopes of getting feedback to raise his bid and still remain below his fellow subcontractors/suppliers). However, in many cases, the subcontractor simply phones or faxes a bid to the general contractors bidding the project and then leaves his office to either work on or visit jobs in progress.

Whatever the situation, general contractors should consider adding an amount to the bid if a low subcontractor/supplier either will not pull or raise his price or if the subcontractor/supplier is not available to discuss his bid.

Chapter 7

AFTER THE BID

POST MORTEM

After the numbers have been given to the contractor's representative who will be submitting the bid proposal, the atmosphere at the contractor's office is filled with a mixture of anticipation and dread.

All of the office staff are hoping that the contractor will be the successful bidder, yet there are lingering, doubting thoughts that no one mentions:
- How much money will be "left on the table" (the difference between the low bidder and second bidder)?
- Did the estimator forget anything?
- Were the numbers added, subtracted, and adjusted correctly?
- Will the project be below the owner's budget?
- Did the person submitting the bid write in the correct numbers on the bid form?

With each telephone call the anticipation grows. When the call finally comes in from the person who submitted the bid, the office will either be celebrating if the contractor is the low bidder or the mood will be somber if the contractor was not successful in the effort.

Due to the high level of stress, mental exertion, and last-minute adjustments, a bid day will cause everyone involved to be thoroughly exhausted. Thus, typically, once the outcome has been determined, those people participating in the bid day activities should be allowed to take the rest of the day off.

If the contractor is not successful in the bid, the estimator should spend the next workday morning reviewing the subcontractor and supplier bids to determine if his bid day analysis was correct. The bids received should also be used to update the database. Upon completion of this post mortem, the estimator should focus his attention and efforts to the next bid that is due.

THE LOW BIDDER

If the contractor has proven to be the successful bidder and the owner makes a decision to award the contract, the contractor has an initial sense of euphoria. As the situation sinks in, the mood shifts to one of caution and partial dread. There are several people on the contractor's staff who have a lot more hard work to be accomplished. The estimator must review the bids received and the estimates for the work to be performed by the contractor very carefully. Each component of the project and the bids received for that component must be analyzed to determine which subcontractor or supplier will be awarded a subcontract or purchase order for each of the construction activities. The estimator should prepare a Bid Analysis and Recap form with each component of the project listed separately. As each bid is analyzed the lowest, responsible bids should be inserted in the appropriate space. (A sample is shown at the end of this section.) This activity is best performed on a computer using a spreadsheet-type of software. A principal of the contractor should become very familiar with the scope of work for the project. If the project manager is not the estimator, the project manager should aid the estimator in analyzing the contractor's bid. The estimator and project manager should also have a coordination meeting to allow the estimator to provide the project manager with all the details of the project and the estimator's intimate knowledge accumulated during the bid preparation.

THE PROJECT MANAGEMENT TEAM

When a general contractor is awarded a construction contract to build or remodel a facility, the company's principals must decide how the project will be managed. There are several popular philosophies used by contractors to manage their projects.

I. **Traditional Method of Managing a Construction Project**

    The traditional manner which is used to manage the project is for management to assign a project manager and a project superintendent to the task of building the project on time and within the contractor's established budget. The project manager will typically handle three to eight projects at a time. The number of projects will depend upon a number of factors: dollar size; complexity; geographical location; and amount of work to be performed by the contractor's own forces. The project manager in this scenario works out of the contractor's home office or regional office. The project superintendent is assigned a single project and works on the job site.

    The contractor may have senior project managers on their staff who supervise three to four project managers.

    Additionally, the type of project might dictate that other personnel be assigned to the job site. These personnel might include the following:

- Assistant superintendent
- Assistant project manager
- Shift superintendents
- Project engineer(s)
- Field engineer(s)
- Project clerk
- Project expediter

For this philosophy, the estimator should include from 4% to 11% of the senior project manager's time in the estimate for each project. Similarly, the estimator should include 12.5% to 33% of the project manager's time in the costs for each project. The full value of all field assigned personnel should be included in the cost of each project.

II. **The "Superman" Approach to Managing Projects**

In this philosophy of managing a project, the roles of estimator, project manager, and project superintendent are assumed by a single individual - a real-life "Superman." An operation's manager supervises the project managers in this concept. Other field support personnel are the same as in the traditional approach.

The benefits of this concept include the following:
- The contractor's overhead is reduced by minimizing the number of estimators required. (See estimating department section.)
- The contractor's weekly payroll is reduced.
- The contractor is more competitive. The cost for the project manager from the traditional approach is deleted.
- The person who estimated the project also performs the buy out and writing of subcontractors and purchase orders. This reduces the potential of items "falling through the cracks."
- The person who estimated the job and prepared the preliminary schedule needed for evaluation purposes in the preparing estimate also the detailed construction estimate.
- The individual who estimated the project is now held accountable for his bid preparation efforts.

THE COORDINATION MEETING

Construction companies that have separate estimating and project management organizations must have the ability to transfer the information prepared by the estimating department to the project management department with a high degree of confidence. Poor communications could result in lost profits. The estimating department has the responsibility of educating the project management department about the project that has been awarded to the construction company. The best method of facilitating this transfer of information is for both parties to dedicate several days to intensive and non-interrupted meetings.

The meeting should include the topics discussed below:

- **Introduction**

    The estimator should provide a brief overview of the scope of work for the project, including any alternates.

- **Review Drawings and Associated Addenda**

    Each drawing should be read and discussed. The project management team will gain much valuable insight from the estimator's point of view.

- **Review Technical Specification and Associated Addenda**

    Each specification should be read and discussed. This discussion will provide the project management team with an excellent idea of the components of the project, as well as how the project should be assembled for buy-out purposes.

- **Review General, Special and Supplemental Conditions, and Associated Addenda**

    Each section of terms and conditions should be read, studied and discussed. The project management team should be able to obtain a feel for the legal restrictions and requirements during this evaluation period.

- **Review of Estimate**

    The estimator should review each component of the estimate with the project management team. This review should include the following items:
    - Review of soils report along with the estimator's interpretation of how the soils report affects costs.
    - Thorough review of components of project which will be performed by the contractor's own forces.
    - Any and all assumptions made during preparation of the estimate.
    - Municipal requirements review.

- **Review of Subcontractor and Supplier Bids Received and Main Summary Sheet**

    The coordination team should review each and every bid received for the project. This will allow the management team to have direct input from the estimating team into how each bid is to be evaluated and interpreted. This activity is an important step in the eventual buy-out process.

- **Review of Initial Schedule**

    The initial schedule should be studied by the coordination team. The project management team should use the initial schedule as a foundation to preparing the project schedule.

- **Review of General Conditions**

    The estimator must discuss the organization which he envisioned to manage the project. The management team should decide who will be required to manage the project after the estimator has given his input.

- **Evaluation of Risk**

    As the coordination meetings progress, both organizations should develop a feel for where the risks are the highest in the project: impossible schedule, electrical subcontractor's bid is too low, no bids were received on the drywall work, etc. The coordination team should make plans to be used in reducing the risks identified.

    The coordination team should not short themselves on the valuable time required to assure a smooth transition. As soon as the meeting is over, each department has its respective new assignments to accomplish.

    The estimating department will rapidly lose interest and intimate knowledge of the project since new challenges and projects are probably waiting while the coordination meeting is taking place. Thus, the management team must obtain as much of the knowledge and information from the estimating team during the coordination meeting as humanly possible.

    Both departments must be aware that they work for the same company and no secrets can go untold if the project is to realize the goal of making the anticipated profit.

Chapter 8

THE PAPERWORK

PROJECT FILES

Each contractor should establish a standard method for setting up and maintaining job site files. Below is a suggested list of files which should be established for each project.

- Bid documents (including all quantity surveys, pricing sheets, quotations, invitation to bid, bid form, and spreadsheets)
- Contract (including original contract and all change orders)
- File for each subcontractor (including subcontract agreement, change orders, correspondence, submittals, bonds, insurance certificate, progress payment billings, copy of checks, copy of lien waivers, copy of supplier lien waivers, preliminary notices, etc.)
- File for each material supplier (including purchase orders, change orders, correspondence, submittals, invoices, copy of checks, preliminary notices, copy of lien waivers, etc.)
- Schedules (including original, weekly look aheads, and updates)
- Correspondence with the owner
- Correspondence with the architect
- Correspondence with financial institution
- Miscellaneous correspondence
- Test reports
- Building permits and inspections
- Submittal log (including transmittals)
- Log of Requests for Information (RFIs)
- Daily reports
- Progress payments
- Time cards
- Cost reports
- Builder's risk insurance
- Punch list(s)
- Warranties and guarantees
- Claims log (including claims)
- Soils report
- Specifications
- Drawings
- As-built drawings

- Close-out
- Meeting minutes
- Progress photos
- Audio cassettes
- Video cassettes
- Project safety
- Minority- and women-owned business reports
- Wage rate reports

These documents should be stored and maintained until the statute of limits expires for the owner to recover for defects. Modern technology allows project records to be scanned and stored on compact disks. This greatly reduces the storage space required.

SCHEDULING

From the owner's standpoint, the most critical standard of performance for the contractor is typically the days taken to complete the project. As such most construction contracts include the statement "time is of the essence." This statement notifies the contractor that a failure to meet the contractual completion time will result in a material breach of the contract, thereby justifying a default termination of the contract.

Both the contractor and the owner reap benefits when a project is completed on or ahead of the scheduled date. The owner's construction financing usually has a higher rate than permanent financing. Permanent financing usually cannot be put into place until the project is complete. Additionally, the project cannot start generating revenue until it is complete. For the contractor, completion in a timely manner precludes extended general condition's costs, labor rate increases, and material increases. Completion of the project also provides the contractor with additional surety credit to pursue other projects.

Due to the importance of time, most owners include a clause in the contract requiring the contractor to submit an initial construction schedule as well as monthly updates.

As a result of the complexity and extent of a construction project, planning and scheduling is a requirement. Planning entails identifying all the activities of a construction project from survey and layout to the punchlist. Scheduling involves correlating the activities by identifying preceding (precedent) and succeeding (successor) activities, assigning a duration to each activity, and graphically representing those activities on a time-scaled diagram to visually indicate the relationships between each activity and the entire construction time period.

This type of scheduling is called the Critical Path Method (CPM). The graphical representation is called a network.

A detailed treatise on the subject of scheduling is beyond the scope of this book. However, a basic understanding is useful and is presented herein.

The most common terms used in networking are defined below.

Activity: any piece of a project which consumes time and resources and has a specific start and end. Activities include items such as survey and layout, submit shop drawings, install door hardware, etc.

Event: the beginning and end point of an activity.

Dummy Activity: an activity that has a time estimate of zero.

Network: a graphical representation of a project plan showing the interrelationships of the project's activities.

Early Start: maximum early finish of preceding activities for an activity.

Early Finish: early start plus activity duration.

Late Finish: minimum of later starts of activities directly succeeding an activity.

Total Float: late start minus early start of an activity or late finish minus early finish of an activity.

Free Float: early start of an activity's successor activity minus early finish of activity.

The early start and early finish times are calculated by performing a forward pass through the network. The late start and late finish times are calculated by performing a backward pass through the network.

A simple example of a network and its calculation is shown below:

EXAMPLE 8.1 CPM Schedule

Activity	Duration	Successor	Predecessor
A	5	--	B, C
B	6	A	D
C	3	A	E
D	2	B	F
E	8	C	F
F	9	D, E	--

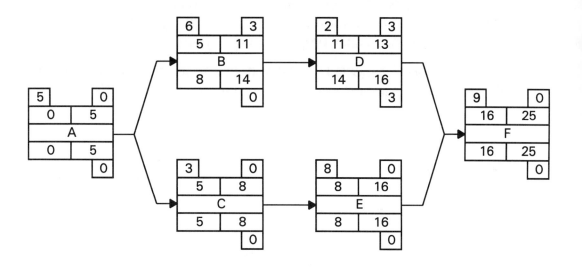

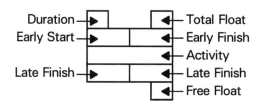

The critical path for the schedule shown above is A - C - E - F.

The free float is the amount of time that the activity completion time can be delayed without affecting the early start of any other activity.

The initial time-scaled CPM network provides the project team (contractor, owner, subcontractors) with the logic and schedule for each activity. As the project progresses, changes to the initial schedule will occur. The critical path, which is defined as the path with the least amount of float, may change as a result of weather delays, material unavailability, strikes, change orders, or a multitude of other reasons.

Thus it is important for the project manager to perform the following tasks:
- Update the schedule regularly.
- Identify delays on the schedule.
- Identify completed activities on an as-built schedule.
- Provide updated schedules to all members of the project team.

Another scheduling tool that should be used on all projects is the two- or three-week "look ahead" schedule. This schedule is prepared by the project superintendent for the purpose of scheduling work at hand and communicating the need for resources to the appropriate entities. The "look ahead" schedule is typically prepared in bar chart format.

Project schedules, especially initial schedules, updated schedules, and as-built schedules are important tools in preparing and proving delay claims.

THE BUY OUT

One of the first tasks to be completed by the project manager is negotiating scope of work and pricing with individual material suppliers and subcontractors, and then writing and releasing subcontracts and purchase orders. This task is commonly referred to as the "buy out."

With respect to dealings with subcontractors and suppliers, the project manager's greatest leverage is realized during the buy out.

The project manager's main goals during buy out are listed below.
- Maximize the anticipated profit for the project.
- Minimize the contractor's risks for the project.
- Shift onerous owner risk shifting clauses to subcontractors and suppliers.

- Ensure that all items of material to be purchased and items of work to be subcontracted are identified, and either purchase orders or subcontracts written and sent out.

If alternates were included in the bid proposal and accepted by the owner, the project manager must evaluate each supplier and subcontractor bid for the effect on award of a contract.

An example of this is shown below:

EXAMPLE 8.2 Effect of Alternates on Subcontractor Selection

The owner requested the following three alternates for the project.

Alternate 1: Install five additional light poles in the parking lot at a location to be determined.

Alternate 2: Add ten isolated grounded receptacles in locations to be determined.

Alternate 3: Add ten 2 x 4 fluorescent lay-in light fixtures in a location to be determined.

The following bids were received from electrical contractors.

Contractor	Base Bid	ALT 1	ALT 2	ALT 3	Total
California Electric	$62,000	$10,000	$900	$1,100	$74,000
Casey Electric	$60,000	$13,000	$1,200	$1,800	$76,000
Continental Electric	$63,000	$11,000	$800	$2,000	$76,800
Columbus Electric	$65,000	$12,000	$1,000	$1,000	$79,000

Since all alternates were accepted, the electrical subcontract, based only on price, should be awarded to California Electric. If only alternates 2 and 3 along with the base bid are accepted, Casey Electric would be the low electrical bidder. The project manager must also examine the exclusions presented by suppliers and subcontractors. For example, if California Electric had excluded temporary power distribution and Casey Electric had included temporary power distribution, valued at $2,500, Casey Electric would be the low electrical subcontractor if all the alternates were accepted.

Most general contractors use standard purchase order and subcontract agreement forms. The purchase order is typically a one-page document indicating the following:

- Vendor's name
- Vendor's address
- Name of project

- Address of project
- Quantities ordered
- Unit prices
- Extension prices
- Taxes
- Terms and conditions of payment
- Shipping arrangements
- Purchase order number
- Contractor's name
- Contractor's address
- Total amount due

Occasionally, the project manager may wish to bind the vendor to a specific reference in the owner/contractor agreement, such as project completion date and liquidated damages. Some contractors use a specific form, a purchase order agreement, for this purpose. Others simply type the language onto the purchase order form. Material purchases that should be considered for this treatment are those that may affect the critical path of the project such as structural steel (materials only), cabinets (materials only), doors and door frames (materials only), window and window frames (materials only), door hardware (materials only), etc.

The contractor/subcontractor agreement is typically a custom-produced document. The subcontract agreement not only references and binds the subcontractor to the owner/contractor agreement but should also address the following issues.

- Subcontractor's scope of work.
- Subcontractor's price to perform the work.
- Subcontractor's specific exclusions.
- Subrogation clause.
- Indemnification clause.
- "Pay-when-paid" clause. (Several courts have set specific guidelines and specific verbiage that must be adhered to in order for this clause to be enforceable.)
- Contract completion date.
- Liquidated damages.
- Date that submittals are required.

As a note of caution, referencing and binding the subcontractor to the owner/contractor agreement without thought could cause problems. As an example, if the owner/contractor agreement contains a bonus clause, that clause should be deleted from the contractor/subcontractor agreement or the contractor may find each subcontractor demanding a piece or all of any bonuses earned by the contractor.

SUBCONTRACTORS

The construction of a project requires numerous and varying skills and trades. These may include concrete work, masonry work, stone work, earthwork, painting, carpentry, electrical, plumbing, flooring, ceilings, HVAC, drywall, plastering, steel work, landscaping, underground utilities, doors, paving, demolition, roofing, insulation, siding, glazing, wallcovering, and a host of other specialty items.

The general contractor's roles and responsibilities are certainly critical in the performance of a construction contract; however, the general contractor typically accomplishes only a small fraction of the actual work in the field.

In today's construction industry there is no one contractor who possesses all the specialized abilities required to complete a project. However, there are many companies in the construction industry that specialize in certain aspects of the business. These specialty firms are called subcontractors.

As a result of the high degree of involvement of subcontractors on a project, the project manager should be aware that subcontractors can and frequently do dictate the success or the failure of a construction undertaking.

The advantages to using subcontractors are listed below.

- The contractor's labor work force is smaller.
- The amount of capital required to construct a project is spread over subcontractors and contractor.
- A subcontractor's firm price has a lower risk than utilizing in-house labor.
- The training required to keep up with materials, standards, codes, and methods is the responsibility of the subcontractor.
- Risks are spread over multiple parties.

The disadvantages of using subcontractors are listed below.

- There is a loss of control of the schedule that is directly proportional to the percent of the project that is subcontracted.
- Subcontractors may not pay their suppliers, thus causing the contractor to pay double.
- Subcontractors may go out of business thus causing schedule and financial distress to the project.
- The introduction of multiple entities complicates the administration of a project.

The selection of subcontractors for a project is an important function of the project manager. The decisions may ultimately determine the general contractor's success or failure of not just the venture but the general contractor's very existence.

When selecting subcontractors, the project manager should, in addition to price, also consider the following factors:

Reputation
- Does the subcontractor perform good workmanship or shoddy workmanship?
- Is the subcontractor oriented toward requesting "nickel and dime" change orders?
- Is the subcontractor dependable?

Financial Stability
- Does the subcontractor have adequate working capital?
- Does the subcontractor have an open line of credit?
- Does the subcontractor have credit with critical suppliers?
- What size of projects (in value) has the subcontractor performed?

Experience
- Is the subcontractor experienced in the type of building being constructed? (Experience with office buildings does not mean the subcontractor understands hospital, grocery store, or clean room construction.)
- Is the subcontractor properly licensed?

Bonding
- Can the subcontractor bond the project?

In dealing with subcontractors, there are several problems inherent to the construction industry that project managers must deal with from time to time. The typical problems are discussed below.

Holding a Subcontractor to Its Bid

Aside from obtaining the lowest and best bids from subcontractors (lowest and best are not necessarily mutually inclusive), the most crucial feature of subcontractor bidding is the ability to hold subcontractors to their bids once submitted. For a general contractor, it is an extremely uncomfortable and nerve-racking experience to be informed that a subcontractor whose bid had been relied upon will not honor that bid.

When a subcontractor refuses to honor its bid, the project manager should follow the guidelines outlined below.

- Send a telegram to the subcontractor requesting confirmation that the subcontractor will sign a contract within twenty-four hours.
- If the subcontractor still refuses to honor its bid, select another subcontractor to perform the work.
- If the contractor suffers financial damage as a result of the new subcontractor selection <u>and</u> if the contractor relied on the first subcontractor's bid (with no modifications) in computing the contractor's bid to the owner, the project manager should weigh the advantages and disadvantages of recovering the loss through litigation. (The doctrine of promissory estoppel is generally used by the courts to base their decisions in this type of case.)

Acceptance of Verbal Bid

A common standard of the construction industry is for a bidding general contractor to receive subcontractor bids on the date that the general contractor's bid is due to the owner. Subcontractor's bids are typically taken over the phone and many are received in the last minutes before the general contractor's bid is due. This puts the general contractor in the position of relying on oral bids, with the potential for substantial financial exposure should any of the subcontractors refuse to honor a bid. Generally speaking, the Statute of Frauds provides that certain types of contracts, including contracts that extend for more than one year and contracts for the sale of goods in excess of $500, are not enforceable unless they are written.

In order to mitigate this situation, the person or persons taking subcontractor bids should request that all subcontracts transmit their bids via facsimile machine right after the verbal quotation is taken.

Flow-Down Clauses

Most general contractors use a customized form of subcontract agreement. However, one clause that is typical to all subcontract agreements is the flow-down clause. This clause contractually ties the subcontractor to the general contractor as the general contractor is tied to the owner. This mutual obligation tends to keep the prime and subcontractor on an even basis, even though there is no privity of contract between the owner and subcontractor.

The project manager has the task to ensure there are no conflicts between the owner/contractor agreement and the contractor/subcontractor agreement. In some cases the project manager may want to exclude some of the clauses from the owner/contractor agreement in the flow-down clause.

An example of this is a bonus clause in the owner/contractor agreement. If this clause is not excluded and the contractor earns a bonus, he might find himself paying every subcontractor the amount of his bonus.

Scope of the Work

The project manager has the task of writing each and every subcontract in such a manner that the scope of work to be included; that the items to be excluded; and that the deliverables are set out clearly, concisely, and are well defined. In writing the scope of work generalizations should be avoided. Exclusions should be kept to an absolute minimum. Listings of work activities and indirect items should include the statement, "The scope of work to be completed by this subcontractor includes, but is not limited to..."

Subcontractor Payments

Most contractor/subcontractor agreements include a "pay-when-paid" clause. This clause makes the general contractor's payment obligation to the subcontractor contingent upon the general contractor's receipt of payment from the owner as an absolute defense of a subcontractor's demand for payment. Some states have taken a stance that is more favorable to general contractors and upheld a subcontractor payment provision whereby the owner's payment to the general contractor is a condition precedent to the general contractor's obligation to pay subcontractors for work that has been performed. However, enforcement of the "pay-when-paid" contractual provisions is in the minority. This amplifies the fact that the contractor should research the owner's financial resources prior to signing a contract. With enforcement of the "pay-when-paid" clause in the minority and an owner who defaults, the contractor will find his payment bond in "open season."

Prior to making payments to subcontractors, the project manager should make an inquiry to suppliers for the subcontractors who have filed preliminary lien notices. The inquiry to those suppliers consists of obtaining information relative to outstanding invoices the supplier has with the subcontractor that can be directly associated with the project.

The project manager must be satisfied that the subcontractor's debt to suppliers will be satisfied with the progress payment. A conditional lien release does not satisfy the need to ensure that the suppliers will be paid. The best way to ensure that the suppliers will be paid is to obtain an unconditional lien release from the suppliers.

As a note, this task of the project manager's is very time consuming and burdensome. However, being caught in the situation of accepting a conditional lien release from the supplier and then having the subcontractor stop payment on the check that is the condition for granting the lien release can cause the general contractor lost profits.

Issuing joint checks is not a cure-all to obtaining unconditional lien releases. If the subcontractor has a dispute with the supplier, the subcontractor is not going to automatically sign the joint check over to the supplier. However, at least the money is still in the general contractor's control until the dispute is resolved.

Change Orders

Whenever an owner requests changes in the work, single and multiple subcontractors will more than likely be affected. As such, it is important that a provision be included in the contractor/subcontractor agreement specifying that work will proceed while the change order is being negotiated. When change orders have been negotiated and agreed upon, the project manager must execute a formal change to the contractor/subcontractor agreement. If the subcontractor is bonded, the general contractor should request an amendment to the bond to cover the change order.

Close-out Documents

The project manager must include a provision in the contract that states the final payment will not be forthcoming until all close-out documents, including but not limited to those listed below, have been received and accepted by the owner:

- Warranties and guarantees
- As-built drawings
- Lien waivers from subcontractor's subs and suppliers

Subcontractor Not Performing

One of the most prevalent problems in dealing with subcontractors is getting them to the job site as scheduled. This is a chronic problem that happens to every project at one time or another.

The first defense for the contractor is to include contract clauses which expressly recognize the right to terminate the contract for default and to recover damages resulting from the default and termination.

The second defense is for the contractor to include contractual language which allows the contractor to take over the subcontractor's purchase orders, sub-subcontracts, and work force in order to keep the flow of work proceeding.

In imposing these clauses the contractor must specifically abide by the notice requirements delineated in the contractor/subcontractor agreement.

In addition to the contract provisions discussed previously, the following areas should also be considered for inclusion of the contractor/subcontractor agreement:

- The subcontractor's responsibility to investigate the site.
- The owner's and general contractor's rights to inspect the subcontractor's work.
- Preparation of daily field reports and submittal to the general contractor.
- A method for handling disputes.
- Project insurance requirements.
- Audit of subcontractor's books.
- Treatment of time (time is of the essence doctrine).
- Not allowing the subcontractor to assign the agreement to other entities.
- Requiring that the subcontractor protect its own work and provide cleanup of its own work.
- Conditions allowing the subcontractor to use contractor's equipment.
- Statement that subcontractor is an independent contractor and must abide by all codes, laws, etc.
- Safety requirements.
- Notice requirements.
- Requiring that subcontractor to layout his own work.
- Workmanship performance requirements.
- Flow-down liquidated damages.
- Not allowing sub-subcontractors without prior approval from the general contractor.

PURCHASE ORDERS

A purchase order is a sheet form of contract normally used to procure materials, equipment rentals, and miscellaneous supplies. It can also be used for procuring professional services such as surveying, soils and concrete testing, and shop drawing preparation. However, the purchase order is not normally used in the case of on-site labor (a subcontract agreement is used to procure on-site labor).

The project manager has the responsibility of writing purchase orders, maintaining purchase order logs and the purchase order numbering system, ensuring that purchased materials are received, approving supplier invoices, and obtaining lien releases.

Purchase orders should include the following information:
- Name of the vendor
- Name of the contractor
- Purchase order number
- Date of the purchase order
- Date the material is required
- F.O.B. point
- Payment terms and conditions
- Item number
- Item quantity
- Unit of measure
- Description of items
- Unit price of item
- Total for the unit
- Sales tax
- Authorized signature for the contractor
- Authorized signature for the supplier's acceptance
- Shipping instructions

As a note, suppliers will not generally accept a purchase order unless the contractor has credit with the supplier. Thus the contractor should set up credit with vendors and suppliers in the area of the project. The contractor should notify the suppliers that only purchase order or cash transactions will be allowed on the contractor's account. The contractor's employees should also be informed that ordering of materials must be through the project manager.

A sample of a purchase order is shown in Fig. 8.1.

ABC Construction Company
222 Main Street
Peakville, Texas 76000
License No. 00234
Telephone No. (555) 686-7000

Purchase Order No. 001
Sheet 1 of 1
Project: Public School

PURCHASE ORDER

The order number must appear on all invoices, packages, packing slips, bills of lading, etc.

To: Sun Lumber Co.
    400 Main Street
    Peakville, Texas 76000

Date: 10/17/93
Terms: Job site
F.O.B.:   Job site
Date Req'd: 10/24/93

Ship to: Public School Project
         300 Main Street
         Peakville, Texas 76000

Item No.	Qty	Unit	Description	Unit Price	Total
1	100	BF	2 x 6 x 10'	.60	$60.00
2	134	BF	2 x 4 x 10'	.60	80.04
3	20	lbs	16d Nails	.20	4.00
Subtotal					$144.04
Sales Tax					8.64
Total					**$152.68**

Accepted By

Authorized By

Figure 8.1 Sample Purchase Order

SUBMITTALS

Throughout the project documents, requirements are specified for the contractor to submit certain information to the owner or architect/ engineer.

The purposes for the submittals are listed below:
- Ensure that the contractor is installing products per the specifications.
- Provide the owner with information relative to the project or project status.
- Provide the owner or architect/engineer with design documents when the specified product is either proprietary or design/build.

Examples of submittals are listed below:
- Original construction schedule
- Updated construction schedules
- Concrete mix designs
- Asphalt mix designs
- Plumbing fixture cut sheets
- Electrical fixture cut sheets
- Reinforcing steel shop drawings
- Structural steel shop drawings
- Project photo documentation
- Fire protection system calculations and drawings
- Door hardware schedule
- Door frame and door shop drawings
- Cabinet shop drawings
- Curtain wall shop drawings
- Test reports
- HVAC equipment cut sheets
- Special doors cut sheets
- Color chips for various products, such as those listed below:
    - Paint
    - Toilet accessories
    - Toilet partitions
    - Metal roofing
    - Prefinished doors

- Samples of various products, such as those listed below:
    - Flooring
    - Wall treatments
    - Ceiling treatments
- Daily logs
- Food service equipment shop drawings
- Elevator equipment cut sheets
- Elevator shop drawings
- Specialty item cut sheets
- Special equipment cut sheets
- Warranties
- Operation and maintenance manuals
- Performance and payment bonds

Many of the submittals require review and sometimes approval of the owner or architect/engineer before an order can be placed for certain products. Thus, if the submission, review, and approval process is not timely, the project schedule may be compromised. This issue is of concern on almost all projects. As such, the contractor must keep and maintain excellent records with regards to submittals.

Recordkeeping for submittals is a tedious, complex, and cumbersome process. However, it should not be taken lightly. One method of maintaining accurate records for submittals is to assign revision number zero to the original submittal.

As each submittal is revised, the revision number would also change. This simple procedure allows project personnel to know the status of a submittal in its history.

One method of submittal recordkeeping is to establish and maintain a submittal log. A sample submittal log is included at the end of this section.

The initial work in preparing a submittal log entails the project manager performing a detailed page-by-page search of the project documents (contract, terms and conditions, specification, drawings, etc.) to find each item which must be submitted. These submittals should be categorized as "For Information Only," "For Review Only," "For Review and Approval," and "For Close-out Only."

The next step involves preparing the submittal log from the information obtained in the discovery phase, as shown in Fig. 8.2.

As items are submitted, returned, and resubmitted, if required, the project manager must update the log. If the submittal process adversely affects the project schedule, the project manager must keep his options open by performing the following tasks.

- Provide the owner with a written notice within the requirements of the contract.
- Update the schedule to show the impact of the submittal process.
- Transmit a copy of the revised schedule to the owner.

In summary, the submittal process is one of the functions for which the project manager is responsible and, at some point, accountable. The submittal log is a double-edged sword. It indicates both the contractor's and the owner's timeliness in submitting, reviewing, and processing key elements of project data.

Project: Public School No. 1
Date: 12/16/93
Revision No.: 4

Figure 8.2 Submittal Log

Submittal Title	Contract Reference	Responsible Party	Revision No.	Date Received	Latest Dates Sent	Returned	Date Forward	Status
CPM Schedule	G.C.	General Contractor	2	N/A	11/13/93			Information
Submittal Schedule	G.C.	General Contractor	3	N/A	11/13/09			Information
Safety Manual	G.C.	General Contractor	0	N/A	9/1/93			Information
Rebar Shop Drawings	3200	Rebar Subcontractor	0	9/10/93	9/12/93	9/17/93	9/17/93	Approved
Concrete Mix Designs	3300	Concrete Supplier	0	9/2/93	9/4/93	9/17/93	9/17/93	Approved
Brux Sample	4200	Masonry Subcontractor	2	10/4/93	10/5/93	10/17/93	10/17/93	Rejected
Structural Steel Shop Drawings	5100	Structural Steel Supplier	1	10/3/93	10/5/93	10/21/93	10/22/93	Resubmit
Cabinet Shop Drawings	6400	Cabinet Supplier	0	11/16/93	11/18/93			Review
Plumbing Fixture Cut Sheets	15100	Plumbing Subcontractor	0	11/17/93	11/18/93			Review
Boiler Cut Sheet	15200	Plumbing Subcontractor	0	11/17/93	11/18/93	12/3/93	12/3/93	Approved

REQUESTS FOR INFORMATION/CLARIFICATION

As a project flows from initiation towards culmination, the design intent becomes more visible. Additionally, during the design phase of the project, it is practically impossible for the architect and engineers to foresee every situation that might arise during construction. Also, theoretical field and design conditions do not always coincide with actual conditions discovered as the project proceeds. There are also questions raised during construction which deal with differences of interpretation.

As a result of the many questions raised and clarifications required the contractor must have a process in place to systematically transmit the questions to the architect and track the response time and change orders or claims resulting from the response. The industry standard for this is to use a Request for Information (RFI) form for submitting the request and an RFI Log for tracking the response time, etc.

The project manager has the responsibility for the RFI process. Outstanding RFIs should be an agenda item during the weekly project meeting with the architect/engineer. RFIs may be in either letter or special document format. The project manager should use a numbering system for whichever format is chosen.

Samples of an RFI Log are shown in Fig. 8.3.

Many claims consultants and attorneys use the RFI log as a tool of great significance in proving delay type claims.

Some owners have already (and others will surely follow suit) incorporated clauses in the contract to reduce the number of RFIs submitted by placing a "fine" on frivolous RFIs. A frivolous RFI is defined as an RFI for which the answer is simply a reference to the plans or specifications with no additional input required to clarify the question. A public agency in California charges contractors $250 for each frivolous RFI.

Figure 8.3 Request For Information Log

Project: _____
Date: _____   **Revision No.** _____

RFI NO.	RFI SUBJECT	DATE SUBMITTED	DATE RETURNED	CHANGE ORDER REQUIRED?

JOB SITE SAFETY

Construction businesses generally devote a great deal of effort into marketing, bidding, scheduling, quality control, cost control, and project management. Safety is normally considered a regulatory requirement that involves spending rather than making money.

The National Safety Council published statistics in 1986 that indicated from 1960 to 1983 the construction industry accounted for more than twice as many disabling injuries and over three and one-half times the accidental deaths of all industries combined.

Accidents cost the construction industry billions of dollars annually. A study conducted by Stanford University and released by the Business Round Table in 1982 estimated that the direct and indirect costs of construction were almost $9 billion (1979 dollars) annually.

Not only does an employer have a legal and moral responsibility to provide a safe working environment, but construction companies should be aware that treating safety as a profit center may reap unbelievable rewards. In the 1990 <u>Engineering News-Record</u> article, "Job site Dangers Defy Worker Protection Drive," Korman reported that a subcontractor was able to reduce its workers' compensation claims from $800,000 to $56,000 per year. Another contractor, who had no formal safety program, paid workers' compensation claims of $500,000 in 1986 after working 600,000 man-hours. In 1987 that contractor hired a safety director and implemented a formal safety program. In 1987 claims were reduced to $300,000 for 700,000 man-hours worked. In 1988 claims were further reduced to $115,000 for 600,000 man-hours. In 1989, more reductions were forthcoming with claims less than $50,000 after 600,000 man-hours.

The keys to improving construction site safety appear to be the following:
- Treat safety as a profit center.
- Establish a formal safety program with input from employees.
- Provide safety training to all employees.
- Stress top management's commitment to safety.

The general contractor must have a written safety program in order to demonstrate management's full commitment to safety. The safety program should address the following:
- Roles and responsibilities with respect to safety for certain individuals listed below:
    - Company safety officer

- Project manager
- Project superintendent
- Operating manager
- General manager
- General superintendent

- Emergency telephone numbers and pager numbers for the following:
    - Police
    - Fire department
    - Hospitals
    - Project manager
    - Owner
    - Project superintendent
    - OSHA
    - Subcontractor management personnel
    - Emergency medical services
    - Utility Companies

- Listing of displays required by state or federal law including the following:
    - Labor standards
    - Safety and health protection on the job
    - Permits
    - Emergency telephone numbers
    - Emergency medical services

- Location of first aid kits on the job site.
- Location of fire protection devices on the job site.
- Description and posting requirements s for signs (i.e., "Hard Hat Area," "Danger," "Caution," "Overhead Electrical Lines," etc.).
- Subcontractor or vendor safety program policies.

- Procedure for excavating near underground utilities.
- Procedure for accident investigations.
- Procedure for handling a job site fire.
- Policy for material safety data sheets (MSDS).
- Procedure for accident reporting and recordkeeping.
- Policy for inspecting the job site for unsafe physical conditions or work practices.
- Housekeeping program.
- Policy for weekly "tailgate" safety meetings

It is the responsibility of the project superintendent to see that the workplace is hazard free and that crew members and subcontractors follow correct safe working habits including wearing of proper protective equipment.

Examples of safety needs to be considered are listed below:
- Covering of vertical rebar ends on which one can fall.
- Installation of a construction passenger elevator for building 60 ft or more in height or 48 ft in depth below ground level.
- Temporary stairs or ladders for building access and exit. If a building or structure is more than three stories or 36 ft, two or more stairways are required.
- Construction areas need be lighted to at least minimum illumination intensities as required by statutes.
- Temporary railings on all open sides of ramps, surfaces, wall openings, floors, or other elevations 7-1/2 ft or more above the ground, floor, or level underneath.
- Elevator shafts that are not enclosed with solid partitions and doors shall be guarded on all open sides by standard railings and toeboards.
- Floor and roof openings through which a worker could fall need to be covered or barricaded.
- Wall openings, from which there is a drop of more than four feet and the bottom of the opening is less than three feet above the working surface need to be guarded.
- Excavations where the bank is 20 ft high or more and the slope is greater than 3/4 horizontal to 1 vertical and when there is

work performed within 10 ft of the edge need be fenced or otherwise guarded.
- A fire protection program for all phases of construction needs to be developed; specific provisions need to be included for portable fire fighting equipment, fixed equipment (including standpipes), and storage of flammable liquids.
- Combustible debris shall be removed promptly during the course of construction. All waste shall be disposed of at intervals determined by the rate of the accumulation and the capacity of the job site container.
- Inspecting all electric and pneumatic tools.
- Maintaining a housekeeping program for the project that is honored by all job site personnel.
- Inspecting stacked materials to ensure that there is no danger of a stack falling over.

JOB SITE HOUSEKEEPING

The benefits of good housekeeping are listed below:
- Saves time and increases construction efficiency.
- Prevents or minimizes construction fires.
- Prevents worker injuries.
- Prevents waste and damage to materials and equipment.
- Provides more room in which to work.
- Improves the quality and quantity of work.
- Eliminates unnecessary delays to other trades.

COST CONTROL

A primary responsibility of the project manager is to assume the role of cost engineer. In this role the project manager performs the following functions:
- Review and approve all invoices applicable to the project.
- Set up the project budget.
- Track the actual costs and compare against budgets.
- Prepare a monthly status report indicating the estimated final cost of the project as compared to original or revised (due to change orders) budgets.

The initial step in preparing a project budget is to convert the estimate worksheet prepared by the estimator into a bid analysis and recap worksheet (see Fig 8.4). Theoretically the total project budget should equal the contract amount with change orders factored into the equation.

After a contract is awarded, the project manager should review the project documents to make an initial list of the individual components (both direct and indirect) that compose the project scope. This list should be converted to a single line spreadsheet (preferably computerized) called the budget worksheet. The costs shown on the bid analysis and recap worksheet should be transferred to the budget worksheet (see Fig. 8.5). For each breakdown item the budgeted costs should be further defined into the following categories:

- Labor: Budget for in house labor to perform the task.
- Material: Budget for items which are buy out and a purchase order is written.
- Subcontract: Budget for work activity which is to be subcontracted.
- Equipment: Budget for equipment associated with an activity, including purchased outright, rented or internally company charged.
- Other Expense: Budget for materials, supplies, expenses and miscellaneous costs associated with an activity which are typically not handled with a purchase order.

Additionally, codes should be established for each cost which the project manager wishes to track. The industry standard is to use a customized form of the Construction Specification Institute numbering system.

For the budget worksheet, each component of the work which is either to be supplied or installed by a separate entity should be shown as a different line item. For example, wood doors might have three separate line items as indicated below.

- Wood doors/material: Line item budget for the wood door supplier.
- Wood doors/labor: Line item budget for either a wood door hanging contractor or for the contractor's own work force.
- Wood doors/other expense: Line item budget to cover buy out items needed to install the wood doors such as hinge templates, carpenter's chisels, shims, etc.

Next, the project manager should review, evaluate and categorize all the bids that were received on bid day. The project manager must select the subcontractors and material suppliers to be used for the project during this evaluation phase.

As the subcontractors, material suppliers, etc. are selected, their base bid plus alternates that have been accepted by the owner should be inserted in the appropriate space of the budget worksheet.

The subsequent step is to analyze the estimate for each portion of the work that is to be performed by the contractor's own work force. The project manager should not blindly insert the estimated costs of force account work into the budget. He should discuss the estimates with the estimator to ensure a good level of confidence in the "numbers."

The project manager must also segregate the contractor's general conditions' costs into appropriate budget levels such as clean-up, survey and layout, field supervision, small tools, etc.

The indirect cost items must next be included in the budget worksheet with the exception of overhead and profit. The next stage is to review the budget worksheet for line items with no budget. After identifying these items, the project manager must estimate their costs and include those costs onto the budget worksheet.

The next activity in the budgeting process is to total the costs on the worksheet and subtract those costs from the contract amount. The remainder from that math problem is the overhead and profit. The project manager should not rush through this process. Additionally, whenever the risk of an item can be mitigated by subcontracting a portion of the work, but the subcontractor's price is higher than the estimate for the contractor to complete that scope of work with in-house labor, the project manager should always consider heavily subcontracting the work unless schedule and quality concerns outweigh the risk of performing the labor.

The budget for a project is a constantly moving target. As items which were left out of the estimate are bought out; as change orders are negotiated; as claims are negotiated; and as items of work are completed, the budget must be updated.

This leads into the process of updating the project budget and forecasting the project costs and overhead and profit on a regular basis.

After setting up the budget, the project manager should utilize a standard format to forecast costs using a spreadsheet with columns entitled "Cost Code," "Activity," "Original Budget," "Change Orders," "Revised Budget," "Cost to Date," "Estimate to Complete," "Estimated Final Cost," and "Variance from Revised Budget" (see Fig. 8.6).

The industry standard is to accrue costs as they occur, but to report accrued costs on a monthly basis. This process is fine for committed costs but has historically been disastrous for force account labor and general conditions types of costs. Committed costs are those for which a subcontract or purchase order has been written and thus the risks of a cost overrun or underrun are slim. In-house labor and materials purchased to support that labor are high risk. Another item that has a potentially high risk associated with it is the general conditions items. These items are generally time based and as such depend mainly on how well the project is proceeding relative to the schedule prepared during the bid process.

Thus, the project manager should track costs for which he has control over more frequently than committed costs. Committed costs can be tracked either monthly or whenever a change to a subcontract or purchase order is made, whichever comes first. Although it may sound cumbersome, costs in the direct control of the project manager should be tracked daily. These costs include the following:

- Field labor for specific and separate construction tasks such as concrete, daily cleanup, rough carpentry, finish carpentry, installation of specialty items (toilet accessories, signs, bulletin boards, etc.), etc.
- Project supervision and administration such as project manager, project superintendent, clerical, project engineer, etc.
- Testing
- Survey and layout
- Rental of trailers, fencing, dumpsters, tools, equipment, etc.
- Utilities, fuel, and telephone
- Job site safety, temporary protection
- Project documentation
- Building permits
- Insurance
- Office supplies, plans and specifications, petty cash, and office equipment
- Bonds

The most difficult cost to track on a daily basis is labor. However, with the help of the project superintendent the task becomes relatively simple. Either at the close of each day or first thing the next morning, the project superintendent should provide the project manager with the daily time card of all in-house personnel who worked on the job site. The time sheet should also indicate the activities worked on by those people.

The project manager should accumulate the cost information (invoices, time cards, and change orders) and distribute the information through the cost report by cost code.

Updating the cost report on a daily basis provides the project manager with an accurate financial picture of the project.

Some of the costs associated with a project must be initially accounted for using a method other than actual costs. These costs include labor burden (benefits, unemployment insurance, FICA, worker's compensation insurance, etc.) and liability insurance. Those costs are typically paid for on a company wide basis and then allocated to the correct job or company overhead. As such, the project manager should use a dependable factor to ensure those costs are shown on the cost report. For example, labor burden might be a separate line item on the cost status report. If company records indicate that the labor burden is 40% of the base labor, the project manager can use the 40% factor applied to the daily labor costs to ensure labor burden is included in the daily costs.

By keeping accurate and timely cost reports, the project manager will greatly reduce the risk of encountering unpleasant surprises concerning the profitability of a project. Samples of the following are included at the end of this section.

- Budget worksheet (Figs. 8.4 and 8.5)
- Cost report (Fig. 8.6)
- Cost codes (Tables 8.1 through 8.17)

Figure 8.4 Bid Analysis And Recap Worksheet

BID ANALYSIS & RECAP WORKSHEET								
PROJECT: Real Estate Office Tenant Improvement								
SPEC NO.	ACTIVITY	LABOR	MATERIALS incl sales tax	SUBS	OTHER incl sales tax	TOTAL TO COMPLETE	MATERIAL BY	LABOR BY
1000	General Conditions	$2,000	$1,000		$1,000	$5,500	GC	GC
6100	Rough Carpentry	1,000	600			1,600	GC	GC
6400	Cabinets				6,000	6,000	RJ	RJ
7200	Insulation				6,114	6,114	ABC	ABC
8100	H.M. Frames	500	1,475			2,075	VALUE	GC
8200	Wood Doors	1,500	3,260			4,960	VALUE	GC
8700	Door Hardware	1,000	2,170			3,370	VALUE	GC
8800	Glass				1,960	1,960	REFLECT	REFLECT
9200	Drywall				13,275	13,275	ANDYS	ANDYS
9300	Ceramic Tile				2,570	2,570	PORTS	PORTS
9500	Acoustical Ceiling				6,110	6,110	JOHNS	JOHNS
9600	Soft Flooring				8,120	8,120	SMITH	SMITH
9900	Painting				6,380	6,380	BRADS	BRADS
15400	Plumbing				6,690	6,690	SIDS	SIDS
15500	Fire Protection				5,435	5,435	HARBOR	HARBOR
15800	H.V.A.C.				7,110	7,110	RAYS	RAYS
16000	Electrical				12,345	12,345	SWIFT	SWIFT
	Profit					12,452	NA	NA
	Bonds					1,681	NA	NA
	Insurance					1,422	NA	NA
	Total	$6,000	$8,505	$0	$83,109	$115,169		

BUDGET WORKSHEET			
PROJECT: Real Estate Office Tenant Improvement			
COST CODE	ACTIVITY	AMOUNT
1030-L	Project Superintendent	$ 2,000
1190-M	Project Documentation	300
1260-M	Printing	200
1510-S	Final Clean-Up	1,000
1570-M	Dumpster	500
1710-O	Liability Insurance	1,422
1810-O	Building Permit	1,500
1820-O	Bonds	1,681
6100-L	Wood Framing - Labor	1,000
6100-M	Wood Framing - Material	600
6200-S	Cabinets	6,000
7200-S	Insulation	6,114
8100-L	H.M. Frames - Labor	500
8100-M	H.M. Frames - Material	1,475
8100-O	H.M. Frames - Other	100
8200-L	Wood Doors - Labor	1,500
8200-M	Wood Doors - Material	3,260
8200-O	Wood Doors - Other	200
8700-L	Door Hardware - Labor	1,000
8700-M	Door Hardware - Material	2,170
8700-O	Door Hardware - Other	200
8800-S	Glass	1,960
9200-S	Drywall	13,275
9300-S	Ceramic Tile	2,570
9500-S	Acoustical Ceilings	6,110
9600-S	Soft Flooring	8,120
9900-S	Painting	6,380
15400-S	Plumbing	6,690
15500-S	Fire Protection	5,435
15800-S	HVAC	7,110
16000-S	Electrical	12,345
	SUBTOTAL	102,717
	PROFIT	12,452
	TOTAL	$115,169

Figure 8.5 Budget Worksheet

Figure 8.6 Cost Report

COST REPORT								
PROJECT: Real Estate Office Tenant Improvement								
REPORT NO. 4								
Cost Code	Activity	Original Budget	Change Orders	Revised Budget	Cost to Date	Estimate to Complete	Estimated Final Cost	Variance	Comments
1030-L	Project Superintendent	$2,000	$150	$2,150	$1,776	$374	$2,150	$0	
1190-M	Project Documentation	300		300	240	60	300	0	
1260-M	Printing	200		200	120	40	160	40	
1510-S	Final Clean-Up	1,000		1,000	0	750	750	250	Bought Out
1570-M	Dumpster	500		500	175	350	525	-25	
1710-O	Liability Insurance	1,422	100	1,522	1,522	0	1,522	0	
1810-O	Building Permit	1,500	150	1,650	1,650	0	1,650	0	
1820-O	Bonds	1,681	175	1,856	1,856	0	1,856	0	
6100-L	Wood Framing - Labor	1,000	600	1,600	1,575	0	1,575	25	
6100-M	Wood Framing - Material	600	400	1,000	880	0	880	120	Bought Out
6200-S	Cabinets	6,000	1,150	7,150	3,250	3,900	7,150	0	
7200-S	Insulation	6,114		6,114	4,500	1,614	6,114	0	
8100-L	H.M. Frames - Labor	500		500	650	-150	500	0	
8100-M	H.M. Frames - Material	1,475		1,475	1,475	0	1,475	0	
8100-O	H.M. Frames - Other	100		100	65	0	65	35	
8200-L	Wood Doors - Labor	1,500		1,500	1,386	114	1,500	0	
8200-M	Wood Doors - Material	3,260		3,260	3,260	0	3,260	0	
8200-O	Wood Doors - Other	200		200	115	85	200	0	
8700-L	Door Hardware - Labor	1,000		1,000	0	1,000	1,000	0	
8700-M	Door Hardware - Material	2,170		2,170	2,170	0	2,170	0	
8700-O	Door Hardware - Other	200		200	0	200	200	0	
8800-S	Glass	1,960	345	2,305	0	2,305	2,305	0	
9200-S	Drywall	13,275	1,100	14,375	12,345	2,030	14,375	0	
9300-S	Ceramic Tile	2,570		2,570	0	2,570	2,570	0	
9500-S	Acoustical Ceilings	6,110		6,110	4,389	1,721	6,110	0	
9600-S	Soft Flooring	8,120		8,120	0	8,120	8,120	0	
9900-S	Painting	6,380	350	6,730	0	6,730	6,730	0	
15400-S	Plumbing	6,690		6,690	5,600	1,090	6,690	0	
15500-S	Fire Protection	5,435		5,435	5,100	335	5,435	0	
15800-S	HVAC	7,110		7,110	6,500	610	7,110	0	
16000-S	Electrical	12,345	986	13,331	11,890	1,441	13,331	0	
	SUBTOTAL	$102,717	$5,506	$108,223	$72,489	$35,289	$107,778	$445	
	PROFIT	12,452	826	13,278	8,894	4,384	13,723		
	TOTAL	$115,169	$6,332	$121,501	$81,383	$39,673	$121,501	$445	

Table 8.1 General Condition Cost Codes

Cost Code	Description
1010	**PROJECT MANAGER** Cost for project manager.
1020	**ASSISTANT PROJECT MANAGER** Cost for assistant project manager.
1030	**PROJECT SUPERINTENDENT** Cost for project superintendent.
1040	**ASSISTANT PROJECT SUPERINTENDENT** Cost for assistant project superintendent.
1050	**PROJECT ENGINEER** Cost for project engineer.
1060	**PROJECT CLERICAL** Cost for project clerical.
1070	**GENERAL PURPOSE LABOR** Cost to the project for general purpose labor.
1100	**INCENTIVE PAY** Bonuses or incentives paid based on job performance
1110	**FICA TAXES** FICA benefit costs on total general condition's staff involved with project.
1111	**FEDERAL AND STATE UNEMPLOYMENT TAXES** Unemployment insurance based on total general condition's staff involved with project.
1112	**WORKER'S COMPENSATION INSURANCE** Workman's Compensation Insurance premiums based on total general condition's staff involved with project.
1113	**MEDICAL INSURANCE** Insurance premiums paid for the general condition's staff involved with project.
1114	**LIFE INSURANCE** Life insurance premiums paid for the general condition's staff involved with project.
1120	**EMPLOYEE MOVING EXPENSES** Costs to relocate project personnel

Table 8.1 General Condition Cost Codes (Cont'd)

1130   TRAVEL EXPENSES
Expenses for meals, travel and lodging associated with the project.

1140   TEMPORARY LIGHTING
Expenses to install and maintain temporary work lights.

1150   PROJECT OFFICE
Trailer rental, repair, modifications, steps into the trailer, security bars, chairs, plan racks, etc.

1160   MOVE IN/MOVE OUT
Job mobilization expenses, delivery, and removal fees if charged.

1170   PROJECT SIGN
The total expense related to installation or procurement of a project identification sign. Does not include any permanent project signage.

1180   OFFICE SUPPLIES
Standard office supplies. For example: note pads, dividers, folders, refreshments, etc.

1190   PROJECT DOCUMENTATION
Expense used to document project progress including photos, videos, cassettes, tape recorders, cameras, etc.

1200   OFFICE EQUIPMENT
Costs for typewriter, computer equipment, printers, fax machine, copy machine, etc.

1210   TEMPORARY STORAGE
Expenses for storage trailers.

1220   PORTABLE TOILETS
Expenses for job site toilets.

1230   TEMPORARY UTILITIES
Expenses for job site utilities including water, power, and gas.

1240   TEMPORARY HEAT/FUEL/FANS
Expenses for temporary heating.

1250   TEMPORARY FENCE
Expenses for job site fencing.

1260   TEMPORARY ROADS AND PARKING
Expenses to provide temporary roads and parking.

Table 8.1 General Condition Cost Codes (Cont'd)

1260   PRINTING
Expenses for blueprints and copies.

1270   POSTAGE AND SHIPPING
Costs incurred from Federal Express or shipping charges for sending or receiving plans, documents, contracts, etc.

1300   SURVEYING
Includes supplies, equipment and contracting expenses.

1310   PROTECTION OF EXISTING/NEW CONDITIONS
Expenses for protecting existing conditions (trees, buildings, etc.) and new conditions (trees, walls, etc.) from damage.

1320   PROJECT VEHICLES
Cost for company owner or leased vehicles associated with the project (includes lease or company rental charge, gas and maintenance).

1430   TESTING/LAB FEES/QC
Quality control services for concrete, steel welds, soil, etc.

1500   GENERAL CLEAN-UP
Expenses for daily project clean-up.

1510   FINAL CLEAN-UP
Final clean-up prior to turning over to owner.

1570   TRASH CONTAINERS/DUMP FEES
Dumpster fees and rental.

1700   BUILDER'S RISK INSURANCE
Expense for builder's risk insurance for the project.

1710   LIABILITY INSURANCE
General ledger transfer cost to apportion liability insurance expenses.

1800   SAFETY-HEALTH-OSHA REQUIREMENTS
Drinking water, ice, salt tablets, first aid supplies, safety hard hats, temporary hand rails, steps and stairs, etc.

1810   BUILDING PERMITS
Cost of all building permits from local city building department.

1820   BONDS
Cost of bonding project.

1900   CRANES
Expenses for general use cranes.

Table 8.1 General Condition Cost Codes (Cont'd)

1910   PERSONNEL/MATERIAL HOISTS
Expenses for vertical hoists.

1920   PARTIES
Expenses for project parties (ground breaking, topping out, etc.)

1930   CLAIMS CONSULTANT FEES
Expense for claims consultant fees associated with the project.

Table 8.2 Site Work Cost Codes

The following codes are to be used if a portion or all of the work described is either subcontracted or performed with in-house forces.

Cost Code	Description
2100	SITE PREPARATION
2200	SITE GRADING, FILLING, AND COMPACTING
2280	STRUCTURAL EXCAVATION AND BACKFILL
2290	REMOVAL OF EXCESS SOIL
2300	DEWATERING
2400	SHORING
2500	SITE UTILITIES (Including water, sewer, and storm systems)
2510	DOMESTIC WATER SYSTEM
2520	UNDERGROUND FIRE LINE
2530	GAS SERVICE
2540	SANITARY SEWER SYSTEM
2550	STORM DRAIN SYSTEM
2600	WOOD PILES
2610	PRECAST CONCRETE PILES
2620	STEEL PILES
2630	DRILLING CAISSON HOLES
2700	ASPHALT PAVING
2710	PAVEMENT MARKINGS
2720	CONCRETE CURBS
2730	CONCRETE GUTTERS

Table 8.2 Site Work Cost Codes(Cont'd)

2740	CONCRETE PAVING
2750	CONCRETE SIDEWALKS
2760	CONCRETE PADS
2770	CONCRETE FOR PIPE BOLLARDS
2780	LIGHT POLE BASES
2800	LANDSCAPING
2810	CHAIN LINK FENCING
2820	WROUGHT IRON FENCING
2830	PLAYGROUND EQUIPMENT
2840	FLAGPOLES
2850	RAILROAD WORK
2870	UNDERSLAB FILL
2900	DEMOLITION AND REMOVALS
2910	ASBESTOS ABATEMENT
2920	LEAD-BASED PAINT ABATEMENT
2930	STORM WATER RETENTION REQUIREMENTS

Table 8.3 Concrete Work Cost Codes

The following codes are to be used if a portion or all of the work described is subcontracted.

Cost Code	Description
3100	CONCRETE WORK
3110	FORMWORK
3120	REINFORCING STEEL
3130	POST TENSIONING
3140	PRECAST CONCRETE

The following codes are used when in-house forces are used to complete all or a portion of the described work.

3300	CONCRETE CAISSONS
3310	PILE CAPS
3320	CONTINUOUS & SPREAD FOOTINGS

Table 8.3 Concrete Work Cost Codes (Cont'd)

3330	GRADE BEAMS
3340	STEM WALLS
3350	CONCRETE WALLS ⟨ 8' HIGH
3360	CONCRETE WALLS ⟩ 8' HIGH
3370	CONCRETE SLABS ON GRADE
3380	CONCRETE SLABS ON DECK
3390	CONCRETE COLUMNS
3400	CONCRETE BEAMS
3410	INTERIOR CONCRETE CURBS
3420	EQUIPMENT PADS
3430	CONCRETE STAIRS
3440	CONCRETE STEPS
3450	STRUCTURAL MATS
3460	ELEVATED SLABS
3470	TILT-UP CONCRETE
3480	GROUTING BASE PLATES

Table 8.4 Masonry Work Cost Codes

The following codes are to be used if a portion or all of the work described is either subcontracted or performed with in-house forces.

Cost Code	Description
4100	MASONRY WORK
4200	MASONRY RESTORATION WORK
4300	MASONRY CLEANING

TABLE 8.5 Steel Work Cost Codes

The following codes are to be used if a portion or all of the work described is either subcontracted or performed with in-house forces.

Cost Code	Description
5100	STRUCTURAL STEEL
5500	ANCHOR BOLTS
5510	CONCRETE EMBEDS
5520	EXPANSION CONTROL
5600	ORNAMENTAL METALS

TABLE 8.6 Carpentry Work Cost Codes

The following codes are to be used if a portion or all of the work described is subcontracted.

Cost Code	Description
6100	WOOD FRAMING

The following codes are used if a portion or all of the work described is either subcontracted or performed by in house forces.

Cost Code	Description
6110	WOOD-FRAMED WALL SYSTEMS
6120	WOOD-FRAMED FLOOR SYSTEMS
6130	WOOD-FRAMED ROOF SYSTEMS
6140	WOOD WALL FURRING
6150	WOOD NAILERS AND BLOCKING
6200	MILLWORK
6210	CABINET WORK
6220	COUNTERTOPS

Table 8.7 Thermal And Moisture Protection Cost Codes

The following codes are to be used if a portion or all of the work described is either subcontracted or performed with in-house forces.

Cost Code	Description
7100	WATERPROOFING AND DAMPPROOFING
7200	INSULATION
7250	FIREPROOFING
7300	SHINGLES AND ROOFING TILES
7400	PREFORMED ROOFING AND SIDING
7500	MEMBRANE ROOFING
7600	FLASHING AND SHEET METAL
7700	ROOF ACCESSORIES
7800	SKYLIGHTS
7900	JOINT SEALERS

Table 8.8 Door And Window Costs Codes

The following codes are to be used if a portion or all of the work described is either subcontracted or performed with in-house forces.

Cost Code	Description
8100	HOLLOW METAL DOORS AND FRAMES
8200	WOOD AND PLASTIC DOORS
8300	SPECIAL DOORS
8400	ENTRANCES AND STORE FRONTS
8500	METAL WINDOWS
8600	WOOD AND PLASTIC WINDOWS
8700	DOOR HARDWARE
8800	GLAZING AND MIRRORS
8900	GLAZING CURTAIN WALLS

Table 8.9 Finish Work Cost Codes

The following codes are to be used if a portion or all of the work described is either subcontracted or performed with in-house forces.

Cost Code	Description
9100	PLASTER WORK
9200	DRYWALL WORK
9300	TILE WORK
9400	TERRAZZO
9500	ACOUSTICAL CEILING WORK
9550	SPECIAL CEILINGS
9600	CARPETING
9620	RESILIENT FLOORING
9650	WOOD FLOORING
9660	SPECIAL FLOORING
9700	EPOXY FLOORING
9800	SPECIAL COATINGS
9900	PAINTING
9950	WALL COVERINGS

Table 8.10 Specialty Work Cost Codes

The following codes are to be used if a portion or all of the work described is either subcontracted or performed with in-house forces.

Cost Code	Description
10100	CHALKBOARDS
10110	TACKBOARDS
10160	TOILET COMPARTMENTS
10185	SHOWER COMPARTMENTS
10210	METAL WALL LOUVERS
10260	WALL AND CORNER GUARDS
10270	ACCESS FLOORING

Table 8.10 Specialty Work Cost Codes (Cont'd)

10305	PREFABRICATED FIREPLACES AND ACCESSORIES
10410	DIRECTORIES
10415	BULLETIN BOARDS
10430	SIGNS
10456	TURNSTILES
10505	METAL LOCKERS
10538	CANOPIES
10551	MAIL CHUTES
10552	MAIL BOXES
10605	WIRE MESH PARTITIONS
10655	ACCORDION PARTITIONS
10675	STORAGE AND SHELVING
10775	TELEPHONE ENCLOSURES
10800	TOILET ACCESSORIES
10880	SCALES

Table 8.11 Equipment Cost Codes

The following codes are to be used if a portion or all of the work described is either subcontracted or performed with in-house forces.

Cost Code	Description
11010	VACUUM EQUIPMENT
11020	SECURITY/VAULT EQUIPMENT
11040	ECCLESIASTICAL EQUIPMENT
11050	LIBRARY EQUIPMENT
11060	THEATER/STAGE EQUIPMENT
11102	BARBER SHOP EQUIPMENT
11104	CASH REGISTER/CHECKING
11106	DISPLAY CASES
11110	LAUNDRY/DRY CLEANING EQUIPMENT
11132	PROJECTION SCREENS
11140	SERVICE STATION EQUIPMENT

Table 8.11 Equipment Cost Codes (Cont'd)

11150	PARKING CONTROL EQUIPMENT
11160	LOADING DOCK EQUIPMENT
11170	WASTE HANDLING EQUIPMENT
11190	DETENTION EQUIPMENT
11400	FOOD SERVICE EQUIPMENT
11458	DISAPPEARING STAIRS
11474	DARKROOM PROCESSING
11476	REVOLVING DARKROOM DOORS
11480	ATHLETIC/RECREATION EQUIPMENT
11488	BOWLING ALLEYS
11496	SHOOTING RANGES
11600	LABORATORY EQUIPMENT
11700	MEDICAL EQUIPMENT
11740	DENTAL EQUIPMENT

Table 8.12 Furnishings Cost Codes

The following codes are to be used if a portion or all of the work described is either subcontracted or performed with in-house forces.

Cost Code	Description
12301	METAL CASEWORK
12350	HOSPITAL CASEWORK
12380	DISPLAY CASEWORK
12390	RESIDENTIAL CASEWORK
12510	WINDOW TREATMENT
12600	FURNITURE AND ACCESSORIES
12690	FLOOR MATS
12700	SEATING

Table 8.13 Special Construction Cost Codes

The following codes are to be used if a portion or all of the work described is either subcontracted or performed with in-house forces.

Cost Code	Description
13010	AIR SUPPORTED STRUCTURES
13250	INTEGRATED CEILINGS
13030	SPECIAL PURPOSE ROOMS
13032	ATHLETIC ROOMS
13034	AUDIOMETRIC ROOMS
13036	CLEAN ROOMS
13038	COLD STORAGE ROOMS
13052	SAUNAS
13054	STEAM BATHS
13081	ACOUSTICAL ENCLOSURES
13091	RADIATION PROTECTION
13120	PRE-ENGINEERED STRUCTURES
13123	GREENHOUSES
13124	PORTABLE BUILDINGS
13160	ICE RINKS
13255	GROUND STORAGE TANKS
13210	ELEVATED STORAGE TANKS
13215	UNDERGROUND STORAGE TANKS
13330	POWER CONTROL SYSTEMS
13650	AIRPORT CONTROL SYSTEMS

Table 8.14 Conveying Systems Cost Codes

The following codes are to be used if a portion or all of the work described is either subcontracted or performed with in-house forces.

Cost Code	Description
14100	DUMBWAITERS
14200	ELEVATORS

Table 8.14 Conveying Systems Cost Codes (Cont'd)

14300	MOVING STAIRS AND WALKS
14400	LIFTS
14500	MATERIAL HANDLING SYSTEMS
14600	HOISTS AND CRANES

Table 8.15 Mechanical Work Cost Codes

The following codes are to be used if a portion or all of the work described is either subcontracted or performed with in-house forces.

Cost Code	Description
15200	PLUMBING
15400	FIRE PROTECTION SYSTEMS
15800	HVAC

Table 8.16 Electrical Work Cost Codes

The following codes are to be used if a portion or all of the work described is either subcontracted or performed with-in house forces.

Cost Code	Description
16000	ELECTRICAL WORK
16100	AUDIO/VIDEO SYSTEM
16200	FIRE ALARM SYSTEM
16300	TELEPHONE SYSTEM
16400	ENERGY MANAGEMENT SYSTEM
16500	CABLE TV SYSTEM
16600	COMPUTER CABLING

PROJECT MEETINGS

During the course of the construction project, the project manager should establish periodic meetings with appropriate entities to discuss submittals, schedule changes, progress payments, and day-to-day problems that are inherent in every construction undertaking. The project meetings are required in order to maintain good working relations with the owner, architect, engineers, subcontractors, bank inspectors, etc., as well as to provide the general contractor's management personnel up to date information on the project's overall status.

A suggested list of meetings and their objectives is indicated below:

I. **Architect/Engineer and Owner's Representative**
- This meeting should be held every week.
- Review the project submittals.
- Review the project schedule.
- Review design problems and changes.
- Review changes requested by the architect/engineer and owner.
- Review claims submitted by the contractor.
- Review overall performances by all parties.

II. **Subcontractors and Suppliers**
- This meeting should be held weekly.
- Review the project schedule.
- Review change requests/claims submitted by each subcontractor or supplier (discussed individually).
- Review the status of submittals.

III. **Progress Payments**
- This meeting should be held monthly.
- Review of progress billings.

IV. **General Contractor's Management**
- This meeting should be held twice a month.
- Review project schedule.
- Review project budget.
- Review potential problems/claims.

- Review subcontractors' performances.
- Review project team member performances.

PROJECT DOCUMENTATION

The relationship of the project team members of a construction venture undergoes several changes during the construction process. The relationships depend mainly on how well or poorly the as-planned project develops into the real concrete, glass, building systems, etc. Other factors affecting those relationships include:
- Quality of workmanship
- Contractor's performance with respect to scheduling deliverables
- Owner's timeliness in making progress payments
- Architect's responsiveness to questions
- Personal relationships developed between the various team member
- representatives
- Performance of subcontractors

The fact that each team member has a different agenda with regards to the project often forces the development of one or more adversarial relationships.

The architect's/engineers' agenda include:
- Maximizing profits for the A/E's.
- Minimizing error's and omissions' insurance claims.
- Transferring risks to the owner and contractor.
- Defining a design that may bring recognition and rewards.

The general contractor's agenda includes:
- Maximizing profits for the general contractor.
- Minimizing construction time.
- Shifting risks to the subcontractors.

The owner's agenda includes:
- Minimizing the project costs.
- Transferring risks to the architect/engineer and general contractor
- Minimizing construction time.
- Being delivered a low maintenance cost building.

The subcontractor's agenda includes:
- Maximizing profits for the subcontractor.
- Minimizing construction time.

The construction lender's agenda includes:
- Maximizing profits for the lender.

This unique situation provides for the high likelihood of the possibility of disputes between the various parties. The prudent and realistic general contractor recognizes the potential for disputes and should design and utilize a project documentation system. A well-implemented documentation system allows the general contractor to be in a dispute-avoidance mode rather than a dispute-resolution mode.

The project documentation system serves the following functions:
- To ensure adequate control and monitoring of the project.
- To document an accurate and complete record of job conditions and problems as well as their overall impact on the project.
- To provide a chronological history of the project.

For those very optimistic contractors the thought of designing and implementing a documentation system for an eye to future disputes may be unpleasant, but failure to adopt such a strategy will assuredly result in disputes. Recognizing, managing, and documenting the problems associated with and encountered on a construction project is burdensome for the contractor's personnel; however, the importance of "building a record" cannot be understated since the entity with the best documentation invariably is successful in a dispute.

Details of the most commonly used types of documentation are provided below. The contractor should also begin collecting facts and documenting the activities of the project as soon as the award is made. Good documentation will aid in estimating disputes.

Progress Photos

One of the most important tools to be put to work in the documentation system is a good camera. Everyone is familiar with the old Chinese proverb that "A picture is worth a thousand words." Every construction project should be equipped with two types of cameras: a 35 mm with wide angle lens, telephoto lens, regular lens, and the ability to place the date on the picture;

and a Polaroid type of camera. The Polaroid allows the person taking the pictures to have the ability to ensure the photo actually depicts the situation.

The 35 mm camera should be used to document the day-to-day progress of the project, as well as backing up the pictures taken with the Polaroid camera. The project superintendent should choose four or five locations and take daily pictures using a wide-angle lens.

The bottom and back of the Polaroid picture allows a place to write in a description of what the picture is depicting and the date. This will result in a routine and systematic pictorial diary of the project.

All the photos should be immediately placed in a photo album. The 35 mm photos should be immediately labeled on the back with a description and date. Additionally the time, location, weather conditions, and personnel shown in the picture should be addressed on the back of the photo. Photographs are particularly useful at documenting defective work and work that is covered up and cannot be viewed at a later date.

Tape Recorder

The contractor's supervisory personnel working on the project should always be in possession of a tape recorder. The project manager usually does not prepare the daily field report so many of the situations encountered by the project manager are not properly documented. As the project superintendent is making hourly rounds of the project, items may be discovered that should be documented and can be recorded at the time of occurrence or discovery.

Each time a new entry is recorded, the user should always state the following information:
- Name
- Company name
- Project name
- Location
- Date and time
- Those who are present
- Weather conditions
- Description of the situation

All project meetings should be either video or tape recorded. If the meetings is tape recorded, each person in attendance should state his name and company name into the tape recorder.

Video Cameras

Each project should also possess a video camera. The video camera can perform all the functions of still photography, only better. It is very important that job site personnel be properly trained in the operation of the video camera. It is also very important that batteries with a full charge are always available.

As with the tape recorder, the operator should always say the following into the microphone for each new video taping:

- Name
- Company name
- Project name
- Location of video
- Date and time
- Identify others in the video
- Weather conditions
- Description of what is being video taped

An important consideration in using the video camera is to remember not to mix or integrate project videos with recreational videos. Project personnel have experienced extremely embarrassing moments when a video was being shown of an important point and then changed to showing the operator involved in a compromising situation.

As with the tape recordings, video tapes should be labeled with the date the tape was full and taken out of the video camera to be taken to storage.

Daily Field Report

From a documentation viewpoint, the most important activity is the preparation of consistent and detailed daily field reports. The daily field report is the foundation for preparing claims to the owner, defending claims from subcontractors, and defending counterclaims from the owner.

Unfortunately most daily field reports lack the information that is useful and are typically considered a nuisance by most project superintendents. It is imperative that the contractor train project superintendents in the importance of properly recording and documenting the daily activities and occurrences of the project.

Daily field reports are generally customized by each contractor. However, as a minimum, daily field reports should contain the following information:

- Name of the project

- Date
- Shift (start time and finish time)
- Name of the person preparing the report
- Weather conditions
- Description of work activities
- Subject of meetings
- Name and number of workers for each subcontractor
- Important deliveries
- Unusual conditions encountered
- Building inspections and results
- Visitors to the job site
- Field directions given to subcontractors
- Verbal directions given by the owner or owner's agent
- Equipment, both used and idle
- Safety issues and accidents

The project manager should also require daily reports from each of the subcontractors working on the project.

Other Documentation

Additional components of the project documentation system include the following:

- Cost accounting records
- Meeting minutes
- Logs (RFI, payment, claims, etc.)
- Speed memos
- Contemporaneous memos
- As-built schedules

PROGRESS BILLINGS AND PAYMENTS

As a construction project continues from start to completion, the general contractor, subcontractors, suppliers, equipment rental businesses, and so forth fund the project from month to month by paying for labor, material, equipment, services, etc., in advance of receiving payment from the owner.

The industry standard is for the general contractor to bill the owner each month for completed work with payment due from ten to thirty days after the owner receives the billing. Contractors then pay their subcontractors and suppliers within fourteen to thirty days who in turn pay their sub-subcontractors and suppliers.

One of the project manager's main functions is to ensure that project billings are submitted promptly and payments from the owner are punctual. The process of billing and receiving payments on a scheduled basis is termed progress billings and progress payments, respectively.

The project manager must complete the actions listed below in order to begin the progress payment process:

- Develop a schedule of values based upon the project budget (see Fig. 8.7).
- Obtain a schedule of values from each of the major subcontractors (see Fig. 8.9).
- Develop a cash flow projection based upon the construction schedule (if required by the contract).

The schedule of values is prepared by itemizing the direct and indirect components of a project and assigning each component a value. Examples of direct components are concrete work, site grading, masonry work, electrical work, and plumbing work. Examples of indirect components include general conditions, overhead and profit, bonds, and insurance. The total of the schedule of values should equal the contract amount.

In developing a schedule of values, an industry practice is to "front end load" the schedule of values. This process consists of applying a higher percentage of the project's overhead and profit to early construction activities. As an example, site preparation may be 12% of the project's costs. The front-end-loading concept would apply greater than 12% of the project's overhead and profit to that line item of the schedule of values.

The practice of front-end-loading can be dangerous and in many cases is abused by contractors. Some contractors have a tendency to redistribute portions of tail-end activities such as painting, acoustical ceiling work, landscaping, asphalt paving, etc., to front-end activities such as concrete work, site preparation, underground utilities, etc. This abuse of front-end-loading provides the contractor with virtual working capital that in reality is a zero-interest loan, which must eventually be paid out to subcontractors and/or suppliers.

Today's construction users are becoming wise to the concept of front-end-loading. Many request that the schedule of values, with overhead and profit, general conditions, bonds, and insurance broken out as separate line items, be submitted with the bid.

Around the twenty-fifth of each month, the project manager should meet with each subcontractor and the project superintendent to prepare a progress billing (application for payment - see Fig. 8.6) to be sent to the owner. The purpose of the meeting is to establish the completion percentage of each activity and to identify materials that have been delivered either to a storage facility or the job site.

With the information obtained from each subcontractor and the project superintendent, the project manager prepares a preliminary project progress billing. The project manager sets up a meeting at the job site with the owner, architect, and financial institution to review the preliminary progress billing. During the meeting, each activity's percent of completion is discussed. For those items that cannot be readily agreed upon, a job walk is performed to view first-hand the work that has been accomplished. At the conclusion of the meeting, the amount to be billed for each schedule of value line item is mutually approved.

The project manager's next responsibility is to prepare and submit the actual application for payment billing. A sample is provided at the end of this section. As part of the progress billing, the project manager must also submit conditional and unconditional lien waivers as required by the contract. The lack of appropriate lien waivers has delayed more payments than all other reasons combined.

One of the owner's duties is to pay the contractor within the parameters established in the terms and conditions of the contract. Many owners delay making progress payments as long as possible to reduce the amount of interest they have to pay on borrowed funds or increase the interest they are getting on their own equity funds used to build the project. The project manager has the responsibility of ensuring the owner makes timely progress payments to the contractor. Additionally, when the owner is late with a payment the project manager should not be shy or procrastinate in the following activities:

- Notifying the owner in writing of the late payment.
- Sending an invoice for interest on the late payment to the owner.
- Filing a lien on the property.

A construction industry practice which is uncommon in other industries is the policy of the owner withholding a portion, ranging from 5% to 20% of the contractor's earnings. This withholding is defined as retention. Industry standard is for retention to be 10% of the billing amount. At the completion of the project the owner has held from 5% to 20% (10% as the norm) of the billings from the contractor. In order to collect the retention, the contractor must complete the punch list; submit all required documentation including as-builts, operation and maintenance manuals, warranties, lien waivers, etc.; provide certificate of occupancy; and complete other contractual obligations. Collection of the retention entails submitting a final invoice to the owner.

Many contractors today are mitigating the effect of retention by negotiating for either lower retention rates, including a contract clause reducing the rate when construction is 50% complete if the project is on schedule or introducing a contract clause allowing subcontractors to collect retention on their work when their obligations are met.

The project manager should keep a history of the progress payment process for documentation purposes. The best way to accomplish this task is to utilize a progress billings and payment log. A sample is shown in Fig. 8.8. The progress billing and payment log is useful in contract law cases involving breach of contract issues.

Figure 8.6 Application For Payment

Project: Real Estate Office Tenant Improvements   Page 1 of 2

Application No.: 4

Period From: 12/1/93 to 12/31/93

Original Contract Sum:	$115,169.00
Change Orders:	6,332.00
Contract Sum to Date:	121,501.00
Total Completed and Stored to Date:	81,758.00
Retainage:	8,175.80
Total Earned to Date:	73,582.20
Previous Payment Applications:	62,145.00
Current Payment Due:	**$11,437.20**

The undersigned contractor certifies that all work included in this application for payment has been accomplished in accordance with the contract documents.

By: _____

SCHEDULE OF VALUES						
APPLICATION FOR PAYMENT APPLICATION NO. 4 PAGE 2 OF 2						
Description of Work	Scheduled Value	Total % Complete	Amount to Date	Stored Materials	Total to Date	Retainage
General Conditions	$ 5,500	39.29%	$ 2,161		$ 2,161	$ 216.10
Rough Carpentry	1,600	100.00%	1,600		1,600	160.00
Cabinets	6,000	54.17%	3,250		3,250	325.00
Insulation	6,114	73.60%	4,500		4,500	450.00
H.M. Frames	2,075	100.00%	2,075		2,075	207.50
Wood Doors	4,960	95.99%	4,761		4,761	476.10
Door Hardware	3,370	0.00%	0	$2170	2,170	217.00
Glass	1,960	0.00%	0		0	0.00
Drywall	13,275	84.71%	11,245		11,245	1124.50
Ceramic Tile	2,570	0.00%	0		0	0.00
Acoustical Ceiling	6,110	71.83%	4,389		4,389	438.90
Soft Flooring	8,120	0.00%	0		0	0.00
Painting	6,380	0.00%	0		0	0.00
Plumbing	6,690	83.71%	5,600		5,600	560.00
Fire Protection	5,435	93.84%	5,100		5,100	510.00
H.V.A.C.	7,110	91.42%	6,500		6,500	650.00
Electrical	12,345	88.33%	10,904		10,904	1,090.40
Profit	12,452	64.79%	8,068		8,068	806.80
Bonds	1,681	100.00%	1,681		1,681	168.10
Insurance	1,422	100.00%	1,422		1,422	142.20
Change Order #1	4,560	100.00%	4,560		4,560	456.00
Change Order #2	1,772	100.00%	1,772		1,772	177.20
Totals	$121,501		$79,588	$2170	$81,758	$8,175.80

Figure 8.7 Schedule Of Values

Figure 8.8 Progress Billing And Payment Log

PROJECT: Real Estate Office Tenant Improvement

DATE: 12/20/93

APPLICATION NO.	DATE SUBMITTED	AMOUNT BILLED	DATE PAID	AMOUNT PAID
1	09/30/93	$11,345.00	10/25/93	$11,345.00
2	10/30/93	$16,156.00	11/26/93	$16,156.00
3	11/30/93	$35,644.00	12/29/93	$35,644.00
4	12/30/93	$11,437.20		

Figure 8.9 Subcontractor Schedule of Values

SWIFT ELECTRIC COMPANY, INC.

PROJECT: Real Estate Office Tenant Improvement

Electrical Rough-In	$5,500.00
Lighting	4,600.00
Outlets and Switched	1,200.00
Clean-Up	410.00
Supervision	635.00
Total	$12,345.00

QUALITY ASSURANCE AND CONTROL

Quality assurance, which involves both quality engineering and quality control, involves 1) the application of standards and procedures to ensure that a product or facility meets or exceeds its desired performance criteria and 2) documentation to verify that the results are obtained.

The quality assurance and quality control objectives of the various parties associated with a construction project differ and are often in conflict. The owner wishes to maximize the quality of characteristics associated with the intended function of the project. The owner, however, also wants to minimize the cost of the project. The architect and engineers seek a level of quality that will assure satisfactory performance of the structure and maintain their reputation as designers, but also to minimize the project's costs. The contractor's interests with respect to quality lie mainly in satisfying the requirements of the plans and specifications at minimum direct cost. Additionally, external influences are increasingly setting quality standards for aspects of the project that may not be directly related to the primary function of the building and do not take into consideration the cost to the owner of such an influence.

From the contractor's point of view, the burden is to meet or exceed the requirement of the specifications, plans, and external influences.

The project manager and project superintendent must familiarize themselves with the quality requirements of the project documents and statutory agencies and establish a system to assure compliance. Ensuring compliance as the work is performed will result in a smaller punchlist and reduction in punchlist and close-out costs.

The project manager should analyze the project documents and prepare a checklist of the quality assurance requirements. The checklist should be given to the project superintendent.

The contractor's quality control program should include the following elements.
- Quality assurance checklist.
- List of testing agencies and their responsibilities.
- Name of surveying and layout company.
- Reports from testing and surveying entities.
- List of inspections required by government agencies.
- Obtaining copies of inspection reports by government agencies.

- Project superintendent's speed memos to subcontractors requesting corrective work.
- Issuance of non-conformance notices to subcontractors procrastinating on making repairs.

The project superintendent who is assigned to the job site on a daily basis has the responsibility of ensuring that the project is built in accordance with the established design criteria. The project superintendent must also ensure that the documentation of the results is forwarded to the project manager.

PROJECT CLOSE-OUT

One aspect of the project that generally involves all parties (contractor, subcontractors, owner, architect/engineer, building officials, financial institution, etc.) is the close-out. The close-out is the final culmination of the efforts (good and bad) required to build the project.

The close-out process concludes the construction phase of the owner/contractor relationship and begins the warranty period. It comes at a time when everyone except the project owner is losing interest in the project. However, the project manager must maintain a focused approach to the close-out process since most owners remember the completion of the project regardless of how successfully or roughly the work was managed.

Closing out the project provides benefits to most of the participants of the project.

- The owner can switch from construction financing to long-term financing.
- The contractor and his subcontractors can collect their retainage.
- The bonding company can remove a contingent liability from its books.
- The construction lender can get repaid.
- The contractor receives additional surety credit to pursue other projects.

Thus it is very important for the project manager to take a proactive approach to project close-out.

The initial responsibility of the project manager involves reviewing the contract documents to identify applicable items that must be included in the close-out package. In Addition, the contractor should have a standard list of items that may not be contractually required, but should be submitted with the close-out (i.e., a letter stating which utilities have been converted to the owner, procedure to be used in making warranty calls, et al.).

A description of close-out items that is typical to most projects is presented below. Whenever a document is sent to the owner or his agent a formal transmittal must accompany that document, with a copy of each going to the close-out file.

Substantial Completion

The project close-out period begins when the contractor believes substantial completion has been attained. Substantial completion is defined as the point at which the project can be occupied by the owner for beneficial use of its intended purpose. Upon agreement of substantial completion, the contractor may not be considered in breach of contract or assessed for further liquidated damages.

The contractor must inform the owner in writing when the contractor feels substantial completion has been achieved. The owner or his agent will then perform a physical inspection of the job site. The inspection will generally produce a punch list. If the items on the punch list are aesthetic in nature, substantial completion will generally be granted. If items appear on the punch list that affect the operation of the building, substantial completion will probably be denied thus keeping the project time clock ticking.

Upon agreement that substantial completion has been attained, the contractor is normally entitled to a significant reduction in retention with a small amount , 1 1/2 to 2 times the value of the punch list work, held until final acceptance.

Punch List

During the substantial completion inspection a single punch list is generated by all parties. This is a list of incomplete items, corrective actions required and incomplete or missing documentation. The term punch list was coined because it was common for owner's agents to punch a hole through the action item with a paper punch once it was completed.

Upon receipt of the punch list the project manager must identify appropriate entities to take action on individual items and to establish a schedule for completion of the punch list.

Guarantees and Warranties

Most construction contracts include a clause on warranty of materials and workmanship. The warranty period is usually one year; however, some owners and state statutes require longer periods. The contractor has the obligation of repairing the items under warranty at no additional cost to the owner.

The term guarantee usually refers to a particular aspect of the work where a longer commitment for repairs is required. Examples of guarantees include roofing systems, elevator equipment, and HVAC equipment. The project manager must obtain guarantees and warranties from all entities who supplied either materials or labor to the project. These documents should be transmitted to the owner as part of the close-out process.

As-Built Drawings

As the project is being constructed, actual versus theoretical/plan locations of systems such as underground plumbing, process piping, electrical, etc., can be determined. Each party responsible for field run systems is required to maintain a record of the location of the system. This record is referred to as as-built drawings. Contract documents typically require the as-built drawings to be submitted at close-out.

Operation, Instruction, and Maintenance Manuals

Operation, instruction, and maintenance manuals should be collected from suppliers and subcontractors by the project manager and transmitted to the owner as a part of the close-out package.

Utilities

The project manager should send a letter to the owner indicating pertinent utility company information (name, type of utility, address, phone number, etc.) and the date the service will be transferred to the owner's name. The project manager should also ensure that deposits are collected from the utility company.

Shop Drawings

Copies of all shop drawings should be transmitted to the owner.

Test Reports

Copies of all test reports should be transmitted to the owner.

Miscellaneous Items

The project manager should transmit a copy of the following miscellaneous items to the owner:
- Shop drawings
- Material cut sheets
- MSDS sheets
- Test reports

Originals of the following items should be transmitted to the owner:
- Certificate of occupancy

- Building inspections
- Special inspections
- Architect's certificate of substantial completion
- Architect's certificate of final completion
- Elevator inspection

Plans and Specifications

Upon completion of the work, the building department approved and "stamped" drawings, specifications, and permits should be transmitted to the owner.

Lien Waivers

One of the important aspects of close-out is furnishing the lien waivers and exchanging releases. Prior to releasing the final payment to the contractor, the owner needs to be sure that its property cannot be subjected to liens from unpaid subcontractors, suppliers, or laborers.

For the general contractor, the project manager must obtain the signature of a supplier's or subcontractor's authorized owner on an unequivocal full lien waiver and release.

The project manager must also secure lien waivers from all subcontractors, suppliers, and independent labor providers who furnished materials or labor to the project.

Stock Materials

The project manager should provide the owner with stock materials such as carpet, tile, paint, ceiling tile, door hardware, etc., as required by the contract.

Project Report Card

When the project is complete, all invoices have been paid, all payments have been received, and all deposits collected, the project manager should prepare a final report card to measure the performance of the project with respect to financial goals, schedule goals, personnel performance, and subcontractor performance. The report should also summarize the relationships between the contractor and owner, architect, engineers, building officials, and financial institutions, respectively.

Chapter 9

CONTRACTUAL CONSIDERATIONS

THE CONTRACT

The section on initial preparation briefly described the sections of the contract and the estimator's role during bidding with respect to the contract. Once the project is awarded, the project manager assumes the role of managing the contract and its provisions.

When two parties agree to a transaction mutually, a contract is formed. In order to reach a point of mutual acceptance of a contract, one party must make an offer and the other must accept the offer.

Solicitation of bids by project owners is not considered as an offer to award a contract. It is an invitation to contractors to submit an offer. The bidding process is simply a way to ensure for the owner that all offers are based upon the same criteria. After the bids have been tendered to the owner, the owner may elect to select one.

Another necessary element for a binding contract is consideration. Consideration is something of value that each party furnishes. Without consideration, there cannot be an enforceable contract. In construction contracting, the contractor offers to provide his services, labor, materials, experience, etc., to construct a physical manifestation. In return, the owner promises to pay a given amount of money.

The project manager has the responsibility of knowing the contract. He must also be aware of the ramifications of either party breaching the contract. With today's litigious society in mind, the project manager must be cognizant and have a good understanding of the legal implications of construction contracting.

Some of the major aspects of a construction contract are discussed below.

Breach of a Contract

A breach of contract is a violation of one or more of the terms, conditions, etc., of the contract by either party. A breach of contract is defined as either a material breach or immaterial breach.

A breach is material if it involves one of the vital aspects of the contract. It is immaterial if it involves a less-important element of the contract. A material breach may justify a default termination of the contract and a lawsuit for damages. An immaterial breach will probably not justify a termination but may entitle the nonbreaching party some form of financial compensation.

Some examples of material and immaterial breaches are listed below in Table 9.1:

Table 9.1 Material And Immaterial Breaches

	Activity	Probable Type	Comments
1	Owner consistently fails to make timely payments.	Material	Default termination is an option.
2	Contractor is two weeks late on a nine month schedule after six months of construction.	Immaterial	Owner may be entitled to damages.
3	Contractor is four weeks late on a nine month schedule after four months of construction.	Material	Default termination is an option.
4	Contractor performs defective workmanship.	Immaterial	Owner has right to require the contractor to repair the defective workmanship.
5	Contractor fails to correct defective workmanship.	Material	Default termination is an option.

As a last note to contractors, material and immaterial breaches of contract are by no means black and white. Before assuming a breach is material, an attorney knowledgeable in construction contracts should be consulted. If a breach is treated as material and is subsequently ruled as immaterial, the originating party will be held in breach and be liable for any damages.

Termination

Termination clauses typically define the following.
- Those shortcomings of the owner that would justify a stop-work by the contractor.

- The owner's right to terminate the contract.

Termination clauses usually include language relevant to the following:
- Termination for default.
- Termination for convenience.

Termination for default is based upon a contractor's material breach of contract. A termination for default may impose onerous demands on the contractor.

Termination for convenience is a right the owner can exercise to terminate the contract for any reason or no reason at all. This type of termination obligates the owner to compensate the contractor.

Damages

Most construction contracts include a clause allowing the owner to recover financial losses as a result of the contractor's failure to complete the project on time. This clause is widely known as a liquidated damages clause.

Since it is difficult to establish or measure the owner's actual damages from a contractor's late completion, the parties establish a liquidated damages amount expressed as a daily charge to be assessed by the contractor for each day that the project is late.

Liquidated damages are generally enforceable as long as they represent a good-faith estimate of the owner's actual financial damages resulting from the contractor's late completion of the project.

A sample calculation of a good faith liquidated damage's sum is shown below:

EXAMPLE 9.1 Liquidated Damages Calculation

The owner has a 10,000 sq ft office building to be leased at $150,000 per year. The financing costs are $600,000 with equity of $100,000. The owner must determine whether the additional finance charges or the loss of revenue will result in greater financial loss. The calculation is shown below:

A. **Loss of Revenue Cost**
   - **Daily loss of revenue** $ 411
   - **Daily cost of architect to continue construction observation**   300

   The total cost as a result of loss of revenue is $711 per day.

B. **Additional Finance Charges**

- Daily financing costs at 10% per year on $700,000 $ 192
- Daily cost of architect to continue construction observation   300

The total cost for additional finance charges is $492 per day.

Thus the owner would include a liquidated damages clause stating the liquidated damages is $711.00 per day.

Liquidated damages have been waived by courts and arbitrators on numerous occasions when there was no formula for how the owner calculated the liquidated damages. Additionally, if the liquidated damages amount is very large and is used as a hammer to intimidate the contractor, those damages will usually not be enforced.

Many contractors believe that if the contract is void of a liquidated damages clause there is no penalty for late completion. This is indeed a false belief. In the absence of a liquidated damages clause, the owner has the right to sue the contractor for its actual or consequential delay damages.

Actual and liquidated damages are not an owner's sole recourse against a tardy contractor. They are a stipulation for delay damages only. The owner may still recover damages for other breaches of contract.

General Conditions

The section on initial preparation describes the general conditions.

Supplementary General Conditions

The section on initial preparation discusses the supplementary general conditions.

Ambiguity

If any provision of the contract is ambiguous, that provision will be interpreted against the party that prepared the contract.

Order of Precedence

Many contracts are complex and contain contradictory statements. The project manager should negotiate an order of precedence agreement with the owner and incorporate the order of precedence into the contract.

The order of precedence establishes a priority for the contract documents, thus allowing contradictory statements to be easily resolved.

A recommended order of precedence is listed below:

- Contract agreement
- Addenda
- Special conditions of the contract
- Supplementary general conditions of the contract
- General conditions of the contract
- Drawings
- Specifications
- Prebid conference meeting minutes
- Other sections

Other important sections of the contract for which the project manager must have a good understanding include the following:

- Change orders
- Time/scheduling
- Progress payments
- Differing site conditions
- Dispute resolution
- Project close-out
- Retention of funds
- Construction law

Although project managers are typically not attorneys, a good understanding of construction law is a requirement for today's project managers.

The law that is applicable in the broadest sense to the design and construction process (i.e., construction law) does not exist as a distinct subject in legal history. It is a melting pot of legal theories developed to be of more general application. However, above all, it is common sense.

In most construction law cases, the question that must be ascertained and resolved between the parties is a very simple one: "What is fair?"

The legal principles of construction law lie in four sources of law:

1. Tort law including the theories of negligence, misrepresentation, fraud, intentional interference, nuisance, intentional harm, fiduciary and other special relationships, defamation, product liability and to the measure of recovery for harm caused.
2. Contract law including a wide range of legal concepts applicable to the creation and enforcement of lawful agreements and to the measure of recovery arising out of breach.

3. Equity law which is based upon basic concepts of fairness and good faith and including theories such as restitution, unjust enrichment, waiver, estoppel, mistake, and the measure of recovery of quantum merit.

4. Statutory law including rights and remedies created by statute such as mechanic's lien statutes, contractor licensing statutes, the Uniform Commercial Code, etc., and statutory constraints which limit, change or supersede common law rights such as statute of frauds, statute of limitations, etc.

With this in mind, the construction contract becomes the benchmark from which construction disputes must be evaluated and measured.

CHANGE ORDERS

The project manager has the responsibility of managing the contractor's change order process. Owner/contractor agreements always include a "changes" clause. This provision allows the owner to make changes and protects the contractor's rights to an equitable adjustment in contract price and schedule. The provision also generally provides a procedural and substantive framework for the final determination of the amount of additive or deductive compensation due to the contractor and the adjustment of time for which the contractor is entitled.

Contract changes may be made pursuant to the terms of the contract. As such most contract forms provide a variety of mechanisms for implementing formal change orders. These mechanisms may include both written and verbal directives. Additionally, most contracts provide for the change work to proceed uninterrupted, even if the parties disagree on the pricing or timing of a change.

When directing the contractor to implement a formal change order, an owner has the prerogative of one of the following actions:

- Requesting a lump sum prior to start of the work.
- Requesting the work to be performed on a time and material basis.
- Requesting a lump sum with the work to begin before the amount is agreed upon thus implementing time and material until the price is agreed upon.

In pricing change orders, the project manager must possess the skills of an estimator. The scope of the change order must be broken down into its components, including direct work, general conditions, and indirect costs. The subcontractors and suppliers affected by a change order must be given a copy of the change request in order to determine the cost for their respective scopes of work. Many contracts limit the overhead and profit percent that both the contractor and subcontractor are allowed on change orders. Some contracts limit the general conditions to a percent of the change order cost. One of the duties of the general contractor is to review supplier and subcontractor change request pricing to ensure that the owner is receiving fair market value for the work.

An example of a formal change order is shown below.

EXAMPLE 9.2 Formal Change Order

The owner has requested the contractor to provide a lump sum to add a wall that is 20 ft long by 8 ft high with the following specifications:

- 3 5/8 metal studs with 5/8" drywall on each side
- 4 electrical outlets on each side
- 1 electrical light switch on each side
- 3 ft x 7 ft door in the middle of the wall with hollow metal frame, solid core stained oak door, 3 hinges, one door closer, one lockset.
- Paint both sides of the wall
- 4 in. resilient base on each side of the wall

The contract allows the general contractor 6% for overhead and profit, 1% for liability insurance, 1% for bond and 5% for general conditions.

The project manager prepared the following for approval by the owner.

Activity	Responsible Party	Amount
Metal Studs & Drywall	ABC Drywall	$ 480.00
Electrical Work	Sparky Electric	440.00
Door Frame - Material	Able Hardware	100.00
Door Frame - Labor	ABC Drywall	30.00
Oak Door - Material	Amco Lumber	160.00
Oak Door - Labor	General Contractor	50.00
Door Hardware - Material	Able Hardware	220.00
Door Hardware - Labor	General Contractor	120.00
Paint Wall	Sun Painting	160.00
Paint Door & Frame	Sun Painting	105.00
Resilient Base	Poly Flooring	40.00

Subtotal	$1,900.00
General Conditions (5%)	95.00
Overhead and Profit (6%)	120.00
Liability Insurance (1%)	22.00
Bond (1%)	<u>22.00</u>
Total Change Order Amount	$2,159.00

As far as impact to schedule, this change probably has none.

If the owner elects to use time and material for the change, the prudent contractor will ensure that all time tickets, material tickets, and purchases are reviewed and signed off by the owner or his agent.

Other types of changes that a contractor must be aware of and train supervisory personnel to understand are:

- De facto Changes
- Constructive Changes
- Cardinal Changes
- Excessive Changes

De facto Changes

Some change provisions provide a method for handling changes that are not of a formal nature, but are the result of oral directives or are necessitated by events or circumstances. The typical manner in which these changes are dealt with is that the owner be notified of the contractor's belief that an oral instruction or event results in a change in the original scope of work. Usually the contractor is required to provide an itemized estimate with supporting calculations and pricing and proposed adjustments in the schedule, all within a contractually established reasonable time period after the event or instruction occurs.

Constructive Changes

Constructive changes occur where, in the face of the contractor's entitlement to a contract change, the owner does not issue a change order or wherever the contractor is forced to bear costs of changed work for which it did not assume the risk. Typically, the contractor's recourse for a constructive change is to file a claim or through litigation.

Cardinal Changes

The typical changes provision in the contract allows the owner to order only changes that fall within the scope of the contract documents. Conduct by the owner which requires the contractor to perform changes which are beyond the scope of the contract documents may constitute a breach of contract.

A cardinal change is one of such character or magnitude that it fundamentally alters the nature or amount of the contractor's work. If the permissible scope of contract work is exceeded, the contract is said to have been abandoned, thus allowing the contractor to recover outside the framework of the contract's change clause.

Examples of cardinal changes include the following:
- Changing the specifications to preclude the contractor from performing the work as bid resulting in magnitude increases in cost.
- Purchase by the owner of adjacent properties and requesting work on the "extended" property.
- Adding stories to a multistory building.

Excessive Changes

The number of changes issued may exceed the scope of the original contract and, as in cardinal changes, result in an abandonment of the contract. Excessive changes usually disrupt the productivity, and scheduled flow, of both changed work and unchanged work. As such a finding of excessive changes may permit use of the total cost method for damage calculations. The contractor's recourse for impact due to excessive changes is typically through the litigation process.

CLAIMS

The construction industry is dispute prone, and as such claims are a fact of life. They are a natural outgrowth of a complex and highly competitive business process during which the unexpected often occurs. For the general contractor the potential for claims cannot and must not be ignored. The best method for dealing with claims is to anticipate them and to avoid them as much as possible. Therefore, contractors must train their project managers in the aspects of claim identification and methods of proving and pricing construction claims.

For the contractor's project personnel, early recognition of a potential claim is necessary to ensure that notice requirements are met and that evidence to support the claim is properly documented and preserved. Thus, familiarity with the contract documents is required in order to recognize claims. All of the contractor's supervisory personnel must possess a working knowledge of the general terms and conditions, plans, specifications, and special conditions of the contract.

Upon determination that a claim merits attention (a claim should not be asserted on every minor incident or disagreement) preparation and organization of the claim should be implemented promptly. All records, documentation, and facts dealing with the claim should be assembled, organized, and reviewed. The person preparing the claim for the contractor should establish a plan for approaching the owner. This plan should include a provision for attempting to settle the issue through negotiations as quickly as possible. Long, drawn-out claims have a tendency to use up dollars and resources that can be put to better use elsewhere.

Along with presentation of the facts to prove the claim, pricing of the claim is another important aspect in being able to settle claims. In pricing claims, there are two approaches available to the project manager. These are the total cost approach and the discrete cost approach.

Total Cost Approach

The total cost approach converts a fixed-price lump sum contract or guaranteed maximum price contract into a pure cost-plus contract. This approach assumes that the events proven by the documentation and personal witnesses resulted in the excess costs above the lump sum or the guaranteed maximum price.

In using the total cost approach the following four tests must be met and proven:

1. The contractor's actual total costs must be correct and reasonable. Thus, the contractor must have in place a cost control system for accurately recording the costs associated with each stand-alone project.
2. The overrun of actual versus original contract lump sum or guaranteed maximum price must be the result of the other party's acts. Costs leading to portions or all of the overrun must not be attributable to the contractor's own inefficiencies, mistakes, mismanagement, or unnecessary work.
3. The original lump sum or guaranteed maximum price must be reasonable and the original pricing properly prepared.
4. The nature of the dispute must be such that it is impossible or impractical to use any other approach.

The total cost approach is not favored by most courts. It basically assumes that the contractor is fault free, which is rarely the case. An example of the total cost approach is shown below:

EXAMPLE 9.3 Total Cost Approach Claim

Contract Costs Incurred
Labor Costs	$1,000,000
Material Costs	250,000
Subcontract Costs	4,300,000
Equipment Costs	175,000
Insurance Costs	100,000
Bond Costs	60,000
Other Expenses	200,000
Overhead and Profit	360,000
Interest on Unpaid Balance	55,000
Claim Preparation Costs	100,000
Total Cost	6,600,000
Revenue Received to Date	5,800,000
Total Claim Amount	$ 800,000

Discrete Cost Approach

Although the discrete costs approach to pricing contractor claims is considerably more complex and complicated than the total cost approach, in most cases, it is generally the best approach. The concept behind the discrete cost approach is to identify each of the events or activities that caused cost overruns and price those separately (discretely). An example is shown below:

EXAMPLE 9.4 Discrete Cost Approach Claim

Loss of Productivity	
200 hours x $35.00 per hour	$7,000
Stacking of Trades	
400 hours x $35.00 per hour	14,000
Extension of schedule	
30 days at $500 per day	15,000
Extension of storm sewer by 200 ft	

Price from utility subcontractor	14,000
Additional paving	
Price from paving subcontractor	<u>35,000</u>
Subtotal	85,000
Overhead and Profit	8,500
Insurance	1,000
Bond	1,000
Interest	5,000
Claim Preparation	<u>8,000</u>
Total Claim	**$108,500**

In preparing claims the project manager must keep in mind the fact that the claim is a sales pitch, with one caveat. The information presented and the monetary request will be audited by agents for the owner. As such, the more factual and forthright the claim, the better the odds are that it will eventually be awarded or settled in favor of the contractor.

DELAYS

The contractor's supervision must be knowledgeable in identifying delays and their impact, both financially and relative to schedule, on the project.

Construction delays are divided into five categories:

1. **Unexcused:** delays caused by the contractor or contractor-controlled entity for which additional costs are absorbed by the contractor (or subcontractor or supplier), result in no time extensions, and render the contractor liable for owner's damages.

2. **Excusable**: delays resulting in extension of time for the contractor but no damages for additional costs.

3. **Compensable**: delays for which the contractor is entitled to be paid costs incurred as a result of the delay.

4. **Concurrent**: delays which are caused by several parties and result in damages being apportioned to the appropriate party.

5. **Sequential**: delay which recognizes an overall delay as caused by different types of delaying events operating sequentially rather than concurrently, and to which the rules of other types of delay will apply for each sequentially.

Unexcused Delays

Examples of unexcused delays are provided below:
- Delay caused by the contractor as a result of the contractor's inability to finance his own work.
- Delay caused by a subcontractor as a result of the subcontractor's failure to finance its own work.
- Failure of contractor to mobilize in a timely manner.

Excusable Delays

Examples of excusable delays are provided below:
- Weather beyond the norm for the area. (The past ten years of weather history is typically used as a point of measurement.)
- Labor strikes.
- Acts of God such as droughts, tornadoes, earthquakes, avalanches, lightning, and flooding.
- Fires.

Compensable Delays

Examples of compensable delays are provided below:
- Failure of owner to allow the contractor timely access to the site.
- Owner restricting access of contractor to the site.
- Failure of owner to provide direction or instructions when clarification is required.
- Failure of owner or his agent to provide timely review and/or approval of shop drawings.
- Failure of owner to supply owner-furnished equipment or materials in accordance with the project schedule.
- Owner-supplied equipment or materials are defective or improper.
- Owner supplying the contractor with inadequate or defective plans and specifications.
- Owner specifying sole source products that are not available.
- Owner or his agent's failure to administrate the contract in good faith.
- Issuance of owner-requested change orders.

- Contractor encountering conditions that were not identified, but will change the scope of work.
- Owner's failure to coordinate other contractors working on the same project.
- Owner's failure to disclose existing construction conditions that would have affected the contractor's price.
- Owner's assumption of risk of loss.

In identifying delay situations, the project manager must be honest in categorizing the type of delay. An unexcused delay as defined between the owner and contractor may also be an unexcused delay between a subcontractor and contractor, thus shifting damages to the subcontractor. In the case of a delay for which the contractor is entitled to damages a claim must be filed with the owner which may result in litigation.

With respect to delays, many owners are including a risk-shifting clause into contracts that provides simply that the contractor's only damage in the event of delay is an extension of time. This troublesome clause for contractors is known as the "no-damage-for-delay" clause. The troublesome part of this clause is that, literally interpreted, the provision requires the contractor to bear the entire cost of a delay caused by the owner. (If this clause is not excluded from the "flow-down" clause found in most subcontract agreements, it could also be interpreted to require the contractor to bear the costs for delays caused by subcontractors.

The potential risk imposed on the contractor by the "no-damage-for-delay" clause is substantial. To the detriment of the contracting industry, many courts have validated the "no-damage-for-delay" clause. On the other hand, some courts and state legislatures have refused to enforce the clause. However, the presence of such a clause in a contract creates a substantial and unacceptable risk to the contractor. As such, the prudent contractor should attempt to negotiate the "no-damage-for-delay" clause out of the contract.

Documenting Delays
Utilizing the Critical Path Method (CPM) scheduling technique to manage the individual work activities of a project has become the industry standard (see the section on scheduling). The CPM provides a baseline for determining the extent of any delays, disruptions, or accelerations to the originally scheduled completion date. That is, if an activity on the critical path is delayed, then the eventual completion date is proportionately affected. In order to prevail with a delay claim, the project manager must establish the length of the delay, the cause of the delay, and the damages as a result of

the delay. In determining the length of a delay, the project manager may adapt one of the approaches described below:

- Total time method: This method determines the delay time by comparing the actual, or as-built, schedule duration to the original, or as-planned, schedule duration. This approach is generally not accepted by judicial bodies since it assumes that the claimant is not responsible for any of the delay.
- Modified total time method: In some cases owners have challenged contractors' original schedules as being too ambitious, too aggressive, or overly optimistic. This usually occurs when a contractor's submitted schedule is shorter than the contract schedule. This method determines the delay time by first modifying the as-submitted schedule duration to represent a more reasonable original schedule (usually to match the contract schedule) and then comparing the as-built schedule duration to the modified as-planned schedule duration.
- Specific occurrence method: This method determines the delay time by first identifying a specific occurrence or event that caused a delay; next, the duration of the specific event is established; thirdly, the event is inserted into the as-planned schedule; and finally, the as-planned schedule duration is compared to the revised as-planned schedule duration. Every event that caused a delay is chronologically inserted into the original as-planned schedule.

NOTICES

Most contracts include a provision that states that the contractor has a certain period of time from the discovery of an issue to give the owner notice of the condition, and to state that the contractor reserves its rights to seek compensation and schedule adjustment upon researching the impact of the issue.

The project manager must begin to immediately document potential claims or disputes upon their discovery. The project manager must also endeavor to comply with any contractual, statutory, or other notice requirements. As such the project manager should maintain a checklist of the notices required in order to ensure compliance.

Although contractors have prevailed with claims despite noncompliance of the written notice requirements, the noncompliance is an unnecessary obstacle that sometimes cannot be overcome.

MECHANIC'S LIENS

A mechanic's lien provides a statutory right for a party to secure payment for work performed and material furnished in the improvement of a private property. When properly executed, a mechanic's lien gives an unpaid contractor, vendor, or workman a security interest in the real estate for which labor and/or materials have been furnished. The purpose of a lien statute is to allow a lien to be perfected upon a premises where a value has been received by the owner, and where the property has been economically enhanced by the furnishing of the labor and/or materials and where the workman, material supplier, and/or contractor have not been paid.

In 1791, Maryland passed the first mechanic's lien law. Since then every state has enacted some form of mechanic's lien law. The project manager must be knowledgeable of the mechanic's lien laws for the geographic location of the project.

Persons or Firms Entitled to Lien Rights

As stated above, lien rights are generally available to any person or firm who provides labor, materials, services, or equipment to the project. For most jurisdictions the following parties are entitled to lien rights:

- Laborers (both skilled and unskilled)
- General contractor
- Specialty contractors (subcontractors)
- Sub-subcontractors
- Material suppliers
- Equipment rental companies

In some jurisdictions parties that provide the services listed below are also entitled to lien rights:

- Architectural design
- Engineering
- Scheduling
- Planning
- Testing and inspection
- Survey and layout

Some jurisdictions limit the maximum amount of a lien to the contract price. Change orders are usually lienable as long as the change is duly authorized per the terms and conditions of the contract. However, claims may or may not be lienable. Interest may or may not be lienable. Delay damages and other damages suffered by the contractor are generally not lienable. Interest may or may not be lienable. Delay damages and other damages suffered by the contractor are generally not lienable.

Requirements of Lien Claimants

All potential lien claimants must meet the requirements listed below:

The material, labor, equipment and/or services must
- have been supplied for the specific improvement or project
- have been incorporated and used for the specified improvement or project
- have become a permanent part of the specific improvement

Due to the fact that liens are considered an extraordinary remedy created by statute, lien claimants must carefully comply with the requirements of the local statute in order to obtain a valid lien. As such, the contractor's project manager must be thoroughly familiar with the requirements for filing a lien and time limitations for notice, filing, and initiation of a foreclosure suit. Typically, time limitations are strictly enforced.

Managing Lien Rights for the Project

The project manager must also monitor the lien rights of subcontractors, material suppliers, subcontractors, subcontractor's suppliers, and professionals such as testing agencies and surveyors.

The project manger must maintain complete and accurate records with regards to lien notices. Most jurisdictions require potential claimants to provide a preliminary lien notice to the owner, the lender, and the general contractor. The general contractor is exempt from filing a preliminary lien notice (see Fig. 9.1) due to the fact that the main purpose of the notice is to make the owner aware that parties other than the general contractor have asserted lien rights on the owner's property. This preliminary lien notice must usually be filed within a specific number of days after services or materials are first provided on the job site. The project manager should check each preliminary lien notice to ensure that the information is accurate and complete.

Claimants have been known to misrepresent the value of work or materials to be performed by either entering no value or entering an exaggerated value on the preliminary lien notice form. The project manager must respond immediately to any errors or omissions on the preliminary notice form. A sample document that can be used to monitor mechanic's lien activities is shown in Fig. 9.1.

With each payment to a subcontractor, supplier, equipment rental company, or any other party with lien rights, the project manager must ensure that the proper lien releases are secured. Lien releases are typically in the two forms described below:

- Unconditional lien release (see Fig. 9.3).
- Conditional lien release (see Fig. 9.4).

As the name indicates, a conditional release requires specific conditions be satisfied before the document legally releases the claimant's lien rights. Owners will generally require the general contractor to provide conditional lien releases with payment requests.

The project manager should never accept conditional lien releases from subcontractor's suppliers, sub-subcontractor's suppliers, or sub-subcontractors. Many general contractors have been the victims of subcontractors who have secured conditional lien releases from their suppliers or subcontractors. The subcontractors either did not pay for the services or materials or placed a stop payment on the check used to pay for the services or materials, thus resulting in the general contractor having to pay the supplier or sub-subcontractor and subsequently trying to recoup the amount from the subcontractor.

Filing a Lien Claim

The project manager must not be bashful or untimely in asserting lien rights when applicable. Asserting mechanic lien rights is relatively easy and inexpensive. In perfecting a claim of lien, the project manager must be careful to adhere to all the statutory technicalities. Specifically, documentation and accounting tasks must be performed. Notice and filing requirements must be strictly adhered to. Generally the claim must be filed and recorded at the County Recorder's office of the county in which the property is located.

After the lien has been filed and the general contractor has not been paid, proceedings must be bought to enforce the lien. The claimant is required to initiate a foreclosure action within a statutory period of time or have its lien dissolved.

Foreclosure of a lien usually results in litigation. The general contractor will incur legal expenses and all the other attendant expenses and delays to settle the issue. The general contractor will be required to prove entitlement to every element of the lien. Usually, if the owner fails to pay the contractor, thus forcing a lien claim, the owner is in serious financial trouble. This results in more than one party being owed monies. Thus, the courts must determine a tier of priorities. This could result in the general contractor receiving only a portion or even none of his claim.

Waiver of Lien

The right to lien may be waived in a number of ways, depending on the local statutes. Description of some of the ways in which lien rights are waived are listed below:

Contractual: Construction contracts sometimes contain a clause whereby the contractor (sub or general) expressly agrees not to file or place any liens against the owner's property.

Statutory: If a potential claimant does not file a claim in accordance with the time requirements or if the claim is inaccurate, the lien rights are generally lost.

The most common method of waiving lien rights is to provide the owner with either a conditional or unconditional lien release upon receipt of monies owed.

In some jurisdictions, the owner may record the contract and payment and performance bond with the County Recorder. This provides the owner with protection from liens filed by subcontractors, material men, equipment rental companies, and labor in direct contract or working for the general contractor. These parties must resort to filing a bond claim if the general contractor does not pay.

In conclusion, mechanic's lien laws are constantly changing. The project manager must stay abreast of lien laws in order to protect the general contractor's interest.

Figure 9.1 Mechanic's Lien Documentation

Name of Claimant: _____

Address: _____
_____

Name of Party to whom claimant is supplying services or materials: _____

Date services first performed or materials received at job site: _____

Preliminary lien notice date: _____

Value of labor and/or material shown on preliminary lien notice: $ _____

Lien Releases

Type*	Date	Amount	Aggregate Total

*conditional or unconditional

Figure 9.2 Preliminary Lien Notice

You are hereby notified that

(name and address of person or firm furnishing labor, services, equipment or material)

has furnished or will furnish labor, services, equipment or materials of the following general description:

for the building, structure, or other work of improvement located at

(address or description of job site sufficient for identification)

The name and address of the person or firm who contracted for the purchase of such labor, services, equipment or material is:

Figure 9.2 Preliminary Lien Notice (Cont'd)

The estimated total price of the labor, services, equipment, or materials furnished or to be furnished is:

$ _____

CONSTRUCTION LENDER	OWNER	ORIGINAL CONTRACTOR
_____	_____	_____
_____	_____	_____
_____	_____	_____

or reputed lender, owner or original contractor

Date: _____

_____
(signature)

_____
(printed name of signed)

_____
(telephone number)

This notice is a statutory prerequisite which established the validity of a potential lien. This notice must be provided to the lender, owner and original contractors not later than _____ after first furnishing labor, materials, equipment, or services.

Figure 9.3 Unconditional Lien Waiver

The undersigned does hereby waive and release any right to a mechanic's lien or stop notice, or any right against a labor and material bond for labor, services, equipment or material furnished to _____

For the job of _____
(owner)

Located at _____

Date: _____   _____
                                                                                            (company name)

By _____

_____
(title)

Figure 9.4 Conditional Lien Waiver

The undersigned is owed the sum of $ _____ on the job of _____ located
            (owner)

for labor, services, equipment or material furnished to _____
_____ through _____
            (date)

Upon payment of said sum to the undersigned, this document shall become effective to release any mechanic's lien, stop notice, or bond right the undersigned has on the above-referenced project through the above referenced date only, but this document does not cover any items furnished (labor, material, or equipment) after said date. If the sum indicated above is paid by check, this release becomes in effect only when said check is paid to the undersigned by the bank upon which it is drawn. Before any recipient of this document relies on it, said party should verify with the undersigned that the sum referred to herein has been received by the undersigned.

Date: _____   _____
                     (company name)

            By _____

Chapter 10

OTHER TOPICS

ERRORS IN THE ESTIMATING PROCESS

One common expression in the construction industry is "The low bidder is the one that either made the most mistakes or left the most out." Contractors snicker and guffaw when that statement is made, but unfortunately there is a lot of truth in that phrase.

Errors during the estimating process have cost construction companies millions of dollars through the years. They have also been one of the main causes of putting contractors out of business.

Many contractors have become bidding machines. They bid eight to ten projects per month without understanding those projects in detail, hoping to cover the scope of work through subcontractor and supplier quotations.

The estimating process is wrought with opportunity for making mistakes and errors. The purpose of this section is to discuss the errors that are prevalent in the estimating process.

Quantity Surveys

During the quantity survey much information is transferred from the drawings to take-off sheets to pricing sheets. Typical errors made during quantity surveys include the following:

- Using the wrong scale to perform measurements of the drawings.
- Transposing numbers from the physical take-off to the quantity collector sheets.
- Missing quantities due to lack of concentration, being disturbed during the take-off, or lack of experience.
- Making simple errors in mathematics (i.e., multiplication, division, addition, and subtraction).
- Using the wrong mensuration formula.
- Using the wrong conversion factor.
- Not allowing for the proper waste.

Pricing the Quantity Survey

Once the quantity survey has been properly allocated to the work components and the quantities for each element tabulated, the estimator must price the labor, material, and equipment for in-house work or apply a rough cost figure to the quantities to arrive at an estimate of a subcontracted value for the work. Typical mistakes made during this phase of the estimate include the following:

- Errors in simple mathematics. (Careless extensions of quantities times unit prices.)
- Using wrong labor rates.
- Using old or unverified material unit prices.
- Transposing figures after extensions are made.
- Inadvertently leaving elements of work off the pricing sheet.

Receiving Subcontractor/Supplier Quotations

Typical mistakes encountered while receiving subcontractor/supplier quotations are listed below.

- Writing the wrong value(s) on the telephone quotation form.
- Writing the wrong scope of work on the telephone quotation form.

Judgment

The estimating process requires many judgment calls. Typical judgment errors are listed below. The errors in judgment listed will probably result in a very competitive, if not successful bid. However, the downside is a reduction in attainable profit for the project.

- Overly optimistic with respect to the time it will take to complete the project, thus resulting in lower than required general conditions.
- Using a subcontractor or supplier bid that is too low.
- Applying overhead and profit that is below the cost of doing business.
- Taking "buy-out" cuts.

Bid Day

"Running" the spreadsheet on bid day is a very nerve-wracking and stressful task. It is especially trying during the last hour before bids are to be submitted. Additionally, providing the person who is actually turning the bid in with the bid form information at the last minute can be heart stopping. Some of the typical bid day errors are listed below:

- Inserting a wrong number onto the spreadsheet.
- Making math errors in cuts or adds on the spreadsheet.
- Showing a cut as an add and vice versa on the spreadsheet.
- Making addition errors to the add/cut columns of the spreadsheet.
- Stating the wrong numbers to the person filling out and turning in the bid proposal form.
- Transposing numbers on the bid proposal form.
- Turning in the bid proposal too late.

COMMON ERRORS IN PROJECT MANAGEMENT

As with the estimating process, the opportunities to make mistakes and errors are plentiful in the project management process. A description of errors typical to project management is provided below.

The Buy Out

Typical buy-out errors include the following:
- Entering the wrong amount on the purchase order or subcontract agreement.
- Not adequately reviewing a subcontractor's scope of work and thus excluding items that the subcontractor should have covered (e.g., not including distribution of temporary power or temporary lighting in the electrical subcontractor's subcontract agreement).

The Schedule

Typical scheduling errors include the following:
- Making mistakes in the logic of the CPM schedule.
- Not updating the schedule as a result of delays or change orders.
- Making judgment errors in estimating durations of tasks.

Notices and Claims

Typical mistakes in the subjects of notices and claims are listed below.
- Providing the owner with a notice of changes conditions, schedule impact, claims, etc., after the time allowed in the contract.
- Not preserving the contractor's rights to additional compensation or time extension by failure to submit a notice.

Progress Payments

Typical mistakes concerning progress payments are described below:
- Submitting untimely progress payments.
- Allowing subcontractors to bill for a higher percentage complete than was actually achieved.
- Underbilling the owner.
- Grossly overbilling the owner.
- Overpaying subcontractors and suppliers.

Other Areas

Miscellaneous mistakes common to projects include the following:
- Failure to maintain proper project documentation.
- Failure to identify changed conditions and claim situations.
- Allowing owner or architect correspondence to go unanswered (silence is acquiescence).
- Failure to keep track of project costs.
- Misinterpretation of contract clauses.

Subcontractors

Typical mistakes/concerns dealing with subcontractors include the following.
- Failure to verify that the subcontractor is properly insured (liability insurance and workman's compensation insurance).
- Failure to put the subcontractor on notice for poor performance (from both a schedule and quality standpoint.
- Failure to obtain the correct lien waivers from subcontractors.
- Failure to provide information to subcontractors (changes to the scope of work, schedule updates, punchlists, etc.).

Many of these errors can be avoided if contractors would adopt and use a meaningful training program.

COMPUTERS IN CONSTRUCTION

The advent of the personal computer during the past decade has brought modern technology to the paperwork-burdened construction industry. The low cost of personal computers and their accessories has allowed even the smallest contractor to computerize portions, if not all, of their operations. Construction processes for which software is readily available are described below.

Estimating

Computers aid estimators in most phases of their work. Activities for which computers may be applied are:

- Preparation, maintenance, and storage of subcontractor and supplier mailing lists.
- Preparation, maintenance, and storage of standardized forms such as:
    - Site investigation checklists
    - Pricing worksheets
    - Scope of work checklists
    - Subcontractor/supplier bid forms
- Preparation, maintenance, and storage of essential databases such as:
    - Labor productivity
    - Labor rates
    - Labor burden factors
    - Material prices
    - Equipment costs
    - Unit prices
- Performance of quantity takeoffs using a digitizer.
- Performing extensions of units of labor, material, equipment, and other expenses to develop the estimate
- Developing spreadsheets for bid preparation.

Project Management

Computer applications for project management functions are listed below:

- "Buy Out" spreadsheets.
- Preparing, updating, and maintaining procurement documents such as subcontractor/contractor agreements and purchase orders.
- Preparation of contractor/owner contractors.
- Preparing, maintaining, updating, and storing project schedules.
- Maintaining the following standardized documents.
    - Mechanic's Lien Release - Conditional
    - Mechanic's Lien Release - Unconditional
    - Warranty
- Maintaining and updating the following information on project subcontractors:

- Liability insurance limits and expiration date.
- Worker's compensation insurance.
- Original contract amount.
- Change orders.
- Maintaining and updating correspondence logs to all project entities.
- Preparing daily field reports.
- Preparing safety records and documentation.
- Preparing and storing RFIs and RFI logs.
- Preparing and maintaining submittal logs.
- Preparing progress payment requests.
- Preparing change order requests.
- Preparing change order requests and claims.
- Preparing, maintaining, and updating project cost reports.
- Maintaining for appropriate use the following standard correspondence:
    - Notice of changed condition to the owner.
    - Reservation of rights for compensation and time extension.
    - Notice to subcontractor of poor performance.
    - Notice to subcontractor to provide cleanup of debris.
    - Notice of claim to owner.
    - Notice of substantial completion.
    - Notice of punchlist completion.
    - Preparing change orders and change order logs.

With scanners and imaging technology, the entire project file can be stored on compact discs rather than in boxes in a storage facility.

As with other technological advances, the general contractor taking advantage of computers must provide training to the personnel using the computers and the appropriate software.

ETHICS IN CONSTRUCTION

The construction industry has the widely held perception of being polluted with rampant dishonesty. Many people outside the industry visualize a con-man when discussing construction personnel. That is an image that champions of the construction industry must change.

The first step to change is acknowledging there is a problem. The second step is to identify the roots of the problem. The final step is to develop an action plan to solve the problem.

A discussion of ethical considerations by topic follows.

Bid Shopping

Bid shopping is the practice of providing one competitor with another competitor's prices before award for the purposes indicated below:
- Subcontractor - realizing a higher profit.
- General Contractor - realizing a higher profit.
- Owner - realizing lower construction costs.
- Architects/Engineers - realizing higher profits.

Bid shopping is practiced in this country by owners, contractors, subcontractors, architects, and engineers. Although it is not illegal in the United States (many commercials by car dealers, computer stores, furniture stores, department stores, etc., state, "We'll match or beat any price in town"), bid shopping is considered unethical in the construction industry.

Contractors who earn a reputation as a bid shopper are usually spurned by knowledgeable subcontractors and suppliers.

Many owners, at the suggestion of the subcontractor industry, require general contractors to list the subcontractors that the general contractor plans to use for a project. The list is usually required to be turned in with the bid. While this practice does tend to reduce bid shopping, it requires the bidding contractor to make snap decisions (a result of the bidding process) that may not be in either the owner's or general contractor's best interest.

Idea Shopping

In order to reduce the construction costs, many owners ask general contractors to submit value engineering ideas after the bids have been submitted. General contractors in turn request value engineering ideas from subcontractors. This process typically turns into just another round of bidding, which results more in subcontractors and general contractors cutting overhead and profit than generating cost-savings ideas.

Additionally, once true cost-savings ideas are presented by one competitor, the user will ask other competitors to provide pricing on the original competitor's idea.

Bid Rigging

Bid rigging is the practice of competitors colluding to defraud owners by staging competitive bids with one of the bidders being predetermined to be the low bidder at a higher-than-normal profit margin. The other bidders tender a higher, but apparently competitive, bid than the predetermined low bidder. This practice is illegal in the United States and has resulted in the convictions of construction company executives.

Substitutions

Substituting is the practice of installing an alternate material to the one specified without the knowledge of the owner or architect/engineer. The main purpose of substituting is to buy the alternate material cheaper than the specified material.

Advance Payments

Contractors (usually unlicensed) who seek and receive advance payments without starting or finishing a project have been one of the main causes of the poor image attributed to the construction industry. One of the best ways for the construction industry to help alleviate this problem is by educating the public on industry standards regarding pay and licensing. This can be accomplished by providing articles to local newspapers.

Overbilling

Overbilling is the practice of charging the owner for work that has not been completed.

Backcharges

Some general contractors commonly, as a course of business, backcharge subcontractors for vague activities at the end of the project in order to increase the overall profit margin. These vague items include cleanup, dumpster charges, repairs, overtime for job superintendent, etc. Subcontractors tend to add these charges to future projects bid to the abusive general contractor.

In conclusion, the best way to avoid conflicts, disputes, tarnished reputations, and claims is to be completely honest and ethical in all business practices. Integrity is the golden rule of business conduct.

TRAINING

One area that most general contractors neglect is training of their key personnel. The construction industry has one of the lowest per capita spending rates for training.

Training provides employees with the following:
- Tools to perform their jobs better.
- Self-confidence in dealing with other people.
- Knowledge to make educated decisions.
- Development of new skills.

Key employees should attend training classes regularly. Additionally, general contractors should sponsor in-house seminars with guest speakers, most of whom would probably be willing to provide their expertise for a nominal fee or none at all.

Table 10.1 is a list of personnel who are instrumental in estimating and project management and the types of regular training they should receive.

Table 10.1 Training Requirements

Estimator	Project Manager	Superintendent
Building Codes	Building Codes	Building Codes
Construction Law	Construction Law	First Aid/CPR
Presentation Skills	Presentation Skills	OSHA
Estimating Techniques	Project Management Techniques	New Construction Techniques
New Construction Products	OSHA Laws and Negotiating Skills	New Construction Products
First Aid/CPR	New Construction Products	Scheduling
New Construction Techniques	New Construction Techniques	Documentation
New Construction Equipment	New Construction Equipment	Scheduling
Scheduling	Preparing Claims	Leadership Skills
Word Processing	Word Processing	Planning Skills
	Scheduling	
	Leadership Skills	

In-house seminar opportunities and potential instructors are listed in Table 10.2 below.

Table 10.2 In-house Seminars

In-House Seminar	Potential Instructor(s)
Bonding	Company Bonding Agent
Insurance	Company Insurance Agent
Safety	OSHA Inspectors
Concrete Formwork Systems	Formwork Salespersons
Asbestos Abatement	Abatement Contractor
Lead Paint Abatement	Abatement Contractor
Elevators	Elevator Subcontractor
Electrical Systems	Electrical Engineer/Subcontractor
Mechanical Systems	Mechanical Engineer/Subcontractor
Curtain Wall Systems	Curtain Wall Salesperson
Building Permit Process	City Building Official
Door Hardware	Hardware Salesperson
Flooring Systems	Flooring Salesperson
Ceiling Systems	Ceiling Salesperson
Drywall and Metal Studs	Drywall and Metal Studs Salesperson
Painting	Painting Salesperson
Sealants	Sealant Salesperson
Specialized Equipment	Equipment Salesperson

Training is also provided by construction industry associations such as:

- Associated General Contractors of America (AGC)
- Association of Builders and Contractors (ABC)
- American Society of Professional Estimators (ASPE)
- American Association of Cost Engineers (AACE)

- Project Management Institute (PMI)

In summary, training and development of employees is a task that should not be overlooked by general contractors. A method to accrue funds for training and development is to add training as a part of labor burden. Budgeting $1,000 per year per employee would add only fifty cents per hour to the man-hour rate.

Chapter 11

MANAGING PROJECT RISKS

The construction industry is an inherently high-risk business. Risks that the project manager faces on a regular basis include the following:

- Subcontractors and suppliers refusing to honor a bid.
- Work that has either been inadvertently left out of the bid or has been underpriced.
- The project becoming financially troubled.
- An employee of a subcontractor being injured on the job site.
- The project owner suspending the project.
- Subcontractors and suppliers installing unauthorized materials.
- Uncommonly bad weather that causes delays and extra costs to the contractor.
- Children playing on the project site during non-working hours.
- Encountering hazardous materials that had not previously been discovered or known.
- Encountering subsurface conditions that differ from those presented by the owner's technical experts
- Refusal of owner to make progress payments
- Lack of timely response from the architect/engineer with regards to the following:
    - Review and approval of shop drawings
    - Clarification requests
    - Information requests
- The owner refusing to respond to the following:
    - Time extension requests
    - Extra compensation requests
    - Information requests
    - Progress payment requests
- Subcontractors and suppliers that become financially distressed.
- Subcontractors and suppliers that refuse to perform the work.
- Subcontractors and suppliers that refuse to honor warranties.

- Accepting onerous risk-shifting clauses in the owner/contractor agreement such as the following:
    - No damage for delay
    - Liquidated damages
    - Indemnification
- Vandalization or thefts on the project site.
- Errors and omissions in the project documents.
- Frivolous mechanics liens and lawsuits filed by subcontractors or suppliers.
- Delays in schedule caused by labor issues (shortage, strikes, etc.).
- Changes in laws that affect the project.
- Acts of God such as earthquakes, floods, tornadoes, etc.
- Poorly produced bid documents.
- Taking on a project that is outside of the company's experience and expertise.

In the business world, risk is typically divided into the following two basic types:

1. Business risk
2. Insurable (pure) risk

Business risk involves the inherent chances that an undertaking will result in either a profit or a loss. A thorough analysis of the business risk associated with a project allows a contractor to weigh the potential for rewards against the potential risk on a given project. The amount of business risk that should be assumed by a contractor is directly proportional to the contractor's tolerance for the risk.

Insurable risk differs from business risk in that insurable risk involves only a chance for loss with no chance for profit.

Business risk and insurable risk can be further divided into the general categories shown below:

Business Risk

A. Economic Risk
- Financial uncertainty
  - Contractor
  - Subcontractors
  - Suppliers
  - Owner
  - Lending institution
  - Design professionals
- Escalation
  - Materials
  - Labor
  - Equipment
- Changes in tax rates

B. Political Risk
- Environmental laws
  - Water
  - Noise
  - Air
- Equal opportunity laws
  - Americans with Disabilities Act
- Government regulations
- Zoning laws
- Permit process

C. Contract Risk
- Risk shifting clauses
  - No damage for delay
  - Indemnification
  - Liquidated damages
- Delays
- Change orders
- Design errors
- Coordination with other entities with direct owner contracts
- Owner/contractor agreement
  - Misinterpretation
  - Misunderstanding
  - Assignment of responsibilities
  - Order of precedence

D. Construction Risk
- Labor
  - Productivity

- Availability
- Skill Level
- Dependability
- Management of
  - Materials
    - Availability
    - Quality
  - Defective Workmanship
  - Equipment
    - Productivity
    - Availability
    - Reliability
  - Construction Sites
    - Access
    - Subsurface conditions
    - Location
- Safety
- Schedule
- Late deliveries
- Subcontractors
  - Liability insurance
  - Worker's compensation insurance
  - Failure to perform
  - Failure to honor warranty
  - Failure to perform
  - Failure to honor bid

D. Management Risk
- Estimating
  - Variation in quantities
  - Variation in productivity
  - Mistakes
  - Omissions
- Bid Proposal
  - Subs/Supplies not honoring bid
  - Mistakes
  - Competition
- Personnel
  - Lack of appropriate experience
  - Lack of appropriate training
  - Competence
  - Staff changes

Insurable (Pure) Risk

A. Natural Disaster (Direct Property Loss)
- Hurricane
- Tornado
- Flood
- Earthquake
- Lightning

B. Man-caused occurrences (direct property loss)
- Fire
- Vandalism
- Theft
- Accidents on job site
- Vehicular accidents (off job site)

C. Personnel Loss
- Personal bodily injury
- Cost to replace employee

D. Indirect Property Loss
- Personal bodily injury

E. Liability Loss
- Public liability loss
- Property damage caused by another party's negligence

A successful contractor must be able to understand the risks and mitigate their effects prior to becoming the victim of the potentially onerous consequences. Specific measures that a contractor may implement to minimize and manage risks are listed below:

- Purchase and maintain adequate liability insurance.
- Purchase builder's risk insurance for every project, whether or not it is an owner requirement.
- Ensure that the contracting firm is listed as an additional insured on all insurance binders of subcontractors and suppliers.
- Ensure that subcontractors and suppliers have appropriate insurance coverage, including, but not limited to, commercial general liability and worker's compensation, and vehicle.
- Require subcontractors to post payment and performance bonds.
  (One study of subcontractors indicated only 14% of subcontractor's work is bonded).
- Negotiate a clause in the owner/contractor agreement giving the contractor the right to suspend work or even terminate the contract as a result of late or nonpayment from the owner.
- Understand the lien laws, especially the notice requirements, of the state in which the project is being contracted.

- Do not waive lien rights.
- Maintain open and honest communications with each member of the project team (owner, designers, subcontractors, suppliers, building officials, lenders, testing agencies, etc.).
- Maintain a file of complete project records, including, but not limited to the following:
    - Estimate and bid documents
    - Project specifications
    - Original plans
    - Revised plans
    - Inspection reports
    - Daily logs
    - Progress photographs
    - Payroll records
    - Schedules and updates
    - Minutes of meetings
    - As-built drawings
    - General correspondence
    - Owner/contractor agreements
    - Change orders
    - Subcontracts
    - Purchase orders
    - Job cost information
    - Progress payment applications
    - Claims
    - Submittals
    - Video record of progress
- Develop and update a CPM schedule for each project.
- Include a backcharge clause in each subcontract agreement.
- Include a comprehensive and legal "pay-when-paid" clause in each subcontract agreement.
- Include a broad form indemnity and hold harmless clause in each subcontract agreement.
- Include a "flow-down" clause in each subcontract agreement.

- Refrain from using bids from subcontractors or suppliers whose prices are too low.
- Study the project soils report.
- Investigate the site during the bidding period. (If allowed, videotape the site visit.)
- Understand the notices clause of the owner/contractor agreement.
- Avoid owners who have a reputation for unfair dealings.
- Avoid architects and engineers who have a reputation of unfair dealings.
- Refrain from entering into contracts with broad form "No-Damage-For-Delay" clauses.
- Refrain from entering into contracts with broad form indemnification and hold harmless agreements.
- Ensure that all agreements made during negotiations are properly documented in the contract.
- Allow adequate time to prepare a bid.
- Do not take on too many bids at any given time.
- Stick with types of projects for which company personnel have experience.
- Examine the cash-flow needs of a project.

The responsible contractor must understand what it hopes to gain or accomplish by accepting a contract and to assess the risks that are associated with achieving its goals. Risk assessment begins by analyzing each and every project that is being considered for onerous consequences. Next, an action plan should be devised to mitigate those risks. The mitigation measures should be compared to the risk to determine if the measure truly minimizes the risk. If it does not, a cost must be associated with the risk. The contractor must determine if the estimated costs for the risk are either offset by potential gains, or should be added to the bid amount. In order to segregate this "cost," contractors may wish to add a line item to the spreadsheet entitled "Risk Factor."

In conclusion, construction companies must do a better job at recognizing and analyzing the risks associated with each and every project as well as business in general.

Chapter 12

LESSONS TO BE LEARNED

CASE STUDIES

One of the best and most respected teaching tools is to study real-life situations. The purpose of this section is to present "real world" experiences with regards to contracting.

I. **Case of the Conditional Lien Waiver**

   The general contractor had a subcontract agreement with the electrical contractor. The electrical subcontractor had a purchase order agreement with a material supplier. Final payment had been made to the general contractor who now owed the electrical contractor retention in the amount of $60,000. The electrical subcontractor had supplied a conditional lien release to the general contractor so the owner would release final payment to the general contractor. The material supplier called the general contractor to inform him that the electrical subcontractor still owed $20,000 on the project. The electrical sub confirmed this statement. The electrical contractor went to the supplier and paid the $20,000 with a check and accepted a conditional lien release. The material supplier phoned the general contractor's bookkeeper relating that he had been paid. When the electrical sub showed up, the bookkeeper released the check without noticing the lien waiver from the supplier was the conditional form. The electrical contractor placed a stop payment on the check given to the material supplier.

   The situation got worse for the general contractor when the supplier liened the project and also placed a claim on the payment bond. In the meantime the electrical contractor closed his shop.

   The end result of accepting the conditional lien release cost the contractor the $20,000 owed to the supplier by the electrical contractor, plus $16,000 in attorney's fees for defending the claims and attempting to collect from the principals of the former electrical subcontractor.

II. Case of the Bid Opening Time

An Arizona-based contractor decided in February to bid a $500,000 job on the Navajo Reservation that had a bid date of April 14, at 2:00 p.m. The prebid conference for the project was held at the site of the project in early March.

On the day of the bid, the contractor dispatched one of his trusted employees to Window Rock, Arizona (Capital of the Navajo Nation) with the bid documents and instructions to call at 1:45 p.m. to obtain the final numbers to be tendered.

The employee carried out his instructions and at 1:50 p.m. had filled out the bid documents and was on his way to the bid opening. Upon his arrival at the bid opening location, he was stunned to discover that the office was locked. He hurriedly searched through the Invitation to Bid (ITB) to make sure that he was at the right location. The ITB confirmed he was, in fact, at the correct address. He then called the main office to inquire if an addendum had changed the bid opening location. The chief estimator told him there were no changes in the location of the bid opening.

What had happened? What went wrong?

The State of Arizona does not observe daylight savings time. However, the Navajo Reservation does, since the reservation covers states other than Arizona. As such, when the contractor attended the prebid conference in early March, the State of Arizona and the Navajo Reservation were in the same time zone. On April 14, the State of Arizona was on Pacific Time, while the Navajo Reservation was on Mountain Time. Thus 2:00 p.m. in Phoenix was 3:00 p.m. in Window Rock. The contractor was one hour late in submitting his bid. The first clue that the contractor had to this situation was the fact that he received no bids after 1:00 p.m. (Phoenix time).

Not only had the contractor lost money by expending labor, phone, and travel expenses to put the bid together, but to add insult to injury, his bid would have been low by less than 2%.

The contractor could have remedied this situation in several different manners. First, obviously, the chief estimator should have paid closer attention to the bid opening time of 2:00 p.m. MDST (Mountain Daylight Savings Time). Second, the contractor should have instructed the person turning in the bid to check in with the owner or the home office upon his arrival in Window Rock.

III. Case of Too Many Bid Proposals

The contractor had decided to bid a $600,000 school job in Chattanooga, Tennessee. Careful examination of the Invitation to Bid indicated that the school district wanted three original bid proposals submitted by each bidding contractor.

On the day of the bid, the contractor dispatched a trusted employee with instructions to call at increments of 1 hour, 45 minutes, 30 minutes and 10 minutes prior to bid time. During the call that was 30 minutes prior to bid time, the employee was given all the information needed on the bid proposals with the exception of the final amount. At the call ten minutes prior to bid time the employee was given the bid amount: $546,000. The bid was turned in one minute prior to the official bid opening time.

The contractor's bid was the first read and as the proposal was read aloud, the employee was horrified to hear that on one of the three bid proposals he had written $564,000, rather than $546,000. The bid opening committee ruled that the official bid from the contractor was $564,000. The other five bidders submitted bids of $551,000, $565,000, $579,000, $591,000, and $610,000. Thus, as a result of filling out one of three bid proposal forms with the numbers reversed, the contractor lost the project.

In this situation, with three proposal forms to submit, the contractor did not give his employee enough time to complete and check the bid documents.

IV. Case of the Half-Size Drawings

The general contractor had just completed a project at a United States Air Force base. The drywall contractor was requesting additional compensation for twice the amount of his subcontract.

During the bidding process the general contractor had ordered half-size drawings to be used in preparing his bid and disseminating to subcontractors to prepare their bids.

As part of the bid preparation, the estimator had priced out the drywall work using unit prices. The estimator had established a price of $59,000 for the drywall work.

The only drywall bid was received from the drywall contractor. His bid was $30,100.

On bid day the estimator called the drywall contractor to discuss the bid since it was so low. The estimator documented the phone call on the back of a telephone quotation form.

During their conversation, the drywall contractor informed the estimator he was aware that the plans were half-size and he had taken that into consideration.

The drywall contractor subsequently signed a subcontract agreement and finished the work. He discovered his error when the job was finished and he discovered twice the material and double the man-hours that had been estimated were consumed.

The drywall contractor's error was in his quantity survey. For each of the drywall areas he used the elevations provided in the plans to perform his take-offs. He made a worksheet which indicated location, length, width, plan area, multiplier, and total area. In the length and width columns he inserted the scaled dimension. He multiplied the scaled length and scaled width together to get his "plan area." The multiplier in all cases was two (his reasoning being that the plans were half-size). The plan area was multiplied by the multiplier to obtain the total area. His worksheet, which was set up to help compensate for the drawings being half-size, had one error. The multiplier should have been four since he was dealing with areas.

The general contractor agreed to present the drywall contractor's case to the Air Force. The claim was denied since it was not presented in a timely manner.

V. Case of the Wrong Subcontract Amount

The general contractor was the low bidder for an Air Force base project. On bid day he had received a bid of $32,850 from a site utilities contractor.

When the project manager wrote the contract he misconstrued the "3" to be an 8 and wrote the subcontract for the amount of $82,850.

The project cost reports showed the amount of award as $32,850. During the course of the project the site utilities contractor submitted three progress payments, each showing a total original contract amount as $32,850.

On the final invoice to the general contractor, the subcontractor changed the original contract amount to $82,850 and billed for the additional $50,000 error plus retainage.

The general contractor refused to pay the $50,000 mistake, even though the president of the company had signed the subcontract agreement.

The site utilities contractor decided to request arbitration in the case. The arbitrator ruled against the site utilities contractor.

This entire affair could have been avoided if any of the following had occurred.
- The site utilities contractor returned the original contract indicating there was an error in the amount.
- The project manager had matched the bid with the contract amount.
- The president empowered the project manager to sign the subcontract agreement. (The project manager would have had another chance to see the subcontract agreement before it was signed and sent to the subcontractor.)

GLOSSARY

acceleration A directive from the project owner which forces the contractor to accelerate a work schedule by completing the project earlier than originally required. Generally, the increased costs caused by acceleration can be recovered by the contractor.

acceptance Assent to the exact terms of the offer, thereby creating a contract.

activity A task or closely related group of tasks whose performance contributes to completion of the overall project.

addendum A written or graphic instrument issued by the architect prior to bidding which modifies or interprets the bidding documents by additions, deletions, clarifications or corrections. An addendum becomes part of the contract documents when the contract is executed.

administrative remedies Resolution of a dispute by a designated administrative tribunal or designated individual. Administrative remedies are used only on public works construction contracts. There is generally a judicial appeal from the administrative decision which is available to dissatisfied contractors.

advertisement for bids Published public notice soliciting bids for a construction project. Most frequently used to conform to legal requirements pertaining to projects to be constructed under public authority, and usually published in newspapers of general circulation in those districts from which the public funds are derived.

agreement The document that formalizes the construction contract. It is the basic contract which states the contract sum and incorporates by reference all of the other documents and makes them a part of the contract.

all risk insurance Insurance against loss arising from any cause other than those perils or causes specifically excluded by name. This contrasts with the ordinary type of policy which names the peril or perils insured against.

allowance A money amount allotted to an item in lieu of an estimated amount.

alternates Additions or subtractions to a contract sum for substitutions asked for by the owner, which the contractor must submit with his proposal.

application for payment Contractor's certified request for payment of amount due for completed portions of work and, if the contract so provides, for materials or equipment delivered and suitably stored pending their incorporation into the work.

approved equal A rather loose term indicating that the contractor may substitute another product for that which the architect specifies, providing it is "equal" in all respects. It is a good idea to accompany a proposal with a list of such substitutions upon which a contractor may have based his figure, because it is the architect, not the contractor, who has the final say as to what is equal.

arbitration A method of dispute resolution whereby the parties to a contract agree, in the contract, to submit any disputes arising out of the contract to binding arbitration. The agreement may prescribe the method of selecting the arbitrators and the rules that will be followed during any arbitration proceeding. Arbitration awards are enforceable in court and may be appealed only on narrow grounds involving fraud or conflict of interest.

as-built drawings See **record drawings**.

award A communication from an owner accepting a bid or negotiated proposal. An award creates legal obligation between the parties.

bar chart   A chart which graphically describes a project consisting of a well-defined collection of tasks or activities, the completion of which marks its end.

bar graph schedule   A graphical schedule relating progress of items of work to a time schedule.

base bid   Amount of money stated in the bid as the sum for which the bidder offers to perform the work, not including that work for which alternate bids are also submitted.

baseball arbitration   A process whereby each party submits a final offer of settlement to a single arbitrator, and the arbitrator, after hearing each party's case, chooses one of the two offers submitted.

beneficial occupancy   Use of a project or portion thereof for the purpose intended.

bid   A proposal stating the sum for which the contractor will complete a project.

bid bond   A bond furnished by a third-party surety guaranteeing that the bidder will honor its bid, if accepted by the owner, and sign the contract at the bid price. If a bidder fails to honor its bid, the owner may recover from the surety the difference between the bid price and the next lowest bid price.

bid date   The date established by the owner or the architect for the receipt of bids.

bid deposit   Monetary deposit required to obtain a set of construction documents and bidding requirements, customarily refunded to bona fide bidders on return of the documents in good condition within a specified time.

bid depository   Under a bid depository plan, subcontractors are required to deposit their bids several hours before prime bid opening time, are restricted from submitting bids outside the depository, and are usually prohibited from withdrawing their bids after the depository has closed. Prime contractors obligate themselves not to accept any bids after the closing of the depository.

bid documents   See **bid package**.

bid form   A form furnished to a bidder to be filled out, signed and submitted as his bid. Also see **bid proposal form**.

bid opening   The opening and tabulation of bids submitted by the prescribed bid time and in conformity with the prescribed procedures.

bid package   All drawings, specifications, documents, estimates, paperwork bid forms, and bid bonds relevant to a construction project. A contract is based on the bid package.

bid price   The sum stated in the bid for which the bidder offers to perform work.

bid proposal form   The form provided to bidding parties for the purpose of submitting bids in a consistent manner. The form may include spaces to be completed by the bidder for lump sum prices, number of addenda, number of days bid is valid, list of subcontractors, days needed to complete the project, pricing for alternates, etc.

bid protest   A bidder's administrative appeal to a public authority complaining that the procurement has been conducted in a manner that violates the laws or regulations governing competitive bidding.

bid rigging   The practice of contractor predetermining the low bidder on future bid projects.

bid schedule   The contractor's calendar of times and dates of projects that the contractor intends to bid.

bid security The deposit of cash, certified check, cashier's check, bank draft, money order, or bid bond submitted with a bid and serving to guarantee to the owner that the bidder, if awarded the contract, will execute such contract in accordance with the bidding requirements and the contract documents.

bid shopping The practice whereby a prime contractor uses the lowest subcontractor bid to induce other subcontractors to reduce their quotes in an effort to obtain the lowest possible price.

bid solicitation An owner's invitation for contractors to submit bids for a construction project. The bid solicitation should contain a complete definition of the work to be performed, as well as a complete statement of the legal terms and conditions that will govern performance of the work.

bid strategy the science (as poker is a science) of outguessing and outmaneuvering the competition in order to secure more profitable work.

bid time The date and hour established by the owner or the architect for the receipt of bids.

bidder One who submits a bid for a prime contract with the owner, as distinct from a sub-bidder who submits a bid to a prime bidder. Technically, a bidder is not a contractor on a specific project until a contract exists between him and the owner.

bidder responsibility A bidder is responsible if it possesses the technical and managerial capability, financial resources, and experience necessary to perform the particular work for which a bid is being submitted. A responsibility determination must be based on a bidder's present condition. Responsibility can be determined after bid opening. If a public project owner reasonably determines that a low bidder is not responsible, the owner need not award the contract to that bidder.

bidding Procedure of formulating the costs (direct and indirect) to construct a project and presenting that cost to the appropriate agency or entity.

bidding documents The advertisement or invitation to bid, instructions to bidders, the bid form, and proposed contract documents including any addenda issued prior to receipt of bids.

bidding period The calendar period beginning at the time of issuance of bidding requirements and contract documents and ending at the prescribed bid time.

bonds Written documents that describe the conditions and obligations relating to the agreement.

breakdown To separate a project into parts; the actual list of parts of a project.

builder's risk fire insurance Protects projects under construction against direct loss due to fire and lightning.

builder's risk insurance A specialized form of property insurance that provides coverage for loss or damage to the work during the course of construction.

building permit A permit issued by appropriate governmental authority allowing construction of a project in accordance with approved drawings and specifications.

buy a job Accept a construction contract at bare cost or below.

buy out The act of procuring services and materials for a construction project.

"cardinal" change Substantial changes outside the general scope of the contract, individually or in cumulative effect, that constitute a breach of contract.

cash allowance Sums set forth by the architect for specific items which the contractor must include in his proposal. Upon completion of a job the contractor must make an accounting for these sums returning any unspent amounts to the owner. Often used for buying hardware, face brick, light fixtures, and other special items.

certificate of insurance A document issued by an authorized representative of an insurance company stating the types, amounts, and effective dates of insurance in force for a designated insured.

certificate of occupancy Document issued by governmental authority certifying that all or a designated portion of a building complies with the provisions of applicable statutes and regulations, and permitting occupancy for its designated use.

certificate of substantial completion A certificate prepared by the Architect on the basis of an inspection stating that the work or a designated portion thereof is substantially complete, which established the date of substantial completion; states the responsibility of the owner and the contractor for security, maintenance, heat, utilities, damage to the work, and insurance; and taxes. The time within which the contractor shall complete the items listed therein.

change order A written change to the drawings and/or specs after the bidding.

changes in the work Changes ordered by the owner within the general scope of the contract, consisting of additions, deletions or other revisions, the contract sum, and the contract time being adjusted accordingly. All such changes in the work shall be authorized by change order, and shall be performed under the applicable conditions of the contract documents. See also **minor changes in the work**.

claim A request by the contractor for a time extension or for additional payment based on the occurrence of an event beyond the contractor's control that has not been covered by a change order.

compensable delay A delay in the contractor's performance of the work that results from the owner's failure to fully and faithfully perform an obligation under the contract. In the absence of a contract provision to the contrary, the contractor is entitled to recover its increased performance costs resulting from the owner-caused delay.

compensatory damages The cost of completing or correcting improperly performed work or the diminished value of the project resulting from improperly performed work.

competitive bidding Contractors submitting lump-sum bids (prices) in competition with other contractors to build the project.

completed operations insurance Liability insurance coverage for injuries to persons or damage to property occurring after an operation is completed (1) when all operations under the contract have been completed or abandoned; or (2) when all operations at one project site are completed; or (3) when the portion of the work out of which the injury or damage arises has been put to its intended use by the person or organization for whom that portion of the work was done. Completed operations insurance does not apply to damage to the completed work itself.

complimentary bid An intentionally high bid submitted as a gesture of good will.

comprehensive general liability insurance A broad form of liability insurance covering claims for bodily injury and property damage that combined under one policy provides coverage for all liability exposures (except those specially excluded) on a blanket basis and automatically covers new and unknown hazards that may develop. Comprehensive liability coverages are available on an optional basis. This policy may also be written to include automobile liability.

concurrent delay A delay resulting from two separate causes at the same time. If excusable or compensable delay occur concurrently with nonexcusable delay, the delay will be treated as nonexcusable.

condition precedent An occurrence that must take place before a party incurs the legal obligation to do something. In the context of construction subcontracting, this issue arises when a subcontract states that the prime contractor's obligation to pay the subcontractor is contingent on the prime receiving payment from the project owner. If the subcontract expressly states that this is a "condition precedent," the payment clause will be enforced according to its literal terms. If the subcontract is less explicit, the clause will be interpreted only to give the prime contractor a reasonable amount of time before the obligation to pay the subcontractor arises.

conditional lien waiver Lien waiver conditioned upon certain event occurring. In most cases the event is a check clearing the bank.

conditions of the contract Those portions of the contract documents that define the rights and responsibilities of the contracting parties and of others involved in the work. The conditions of the contract include general conditions, supplementary conditions, and other conditions.

confidential listener A person whom two disputing parties employ to hear their confidential settlement proposals and inform them if the proposals are within a certain percentage of each other.

consent of surety Written consent of the surety on a performance bond and/or labor and material payment bond to such contract changes as change orders or reductions in the contractor's retainage, or to final payment, or to waiving notification of contract changes. The term is also used with respect to an extension of a time in a bid bond.

consequential damages Indirect losses caused by breach of the contract. These damages are recoverable if they were reasonably foreseeable at the time the contract was formed.

consequential loss Loss not directly caused by damage to property, but which may arise as a result of such damage, i.e., damage to other portions of a building or its contents due to roof leaks.

consideration Something of value. Some form of consideration must be furnished by each party to the contract. Consideration may be a promise to perform in the future or a promise to pay.

construction cost (for calculating compensation to the architect): The total cost or estimated cost to the owner of all elements of the project designed or specified by the architect, including at current market rates (with reasonable allowance for overhead and profit) the cost of labor and materials furnished by the owner and any equipment that has been designed, specified, selected, or specially provided for by the architect, but not including the compensation of the architect and the architect's consultants, the cost of land, rights-of-way, or other costs which are the responsibility of the owner.

construction documents Drawings and specifications setting forth in detail the requirements for the construction of the project.

construction management A method of organizing a construction project where the owner awards multiple prime contracts to various trade contractors. The project is usually phased, with construction starting on early phases before final working drawings have been completed for the later phases of the project.

constructive change   A change in the work that the contractor alleges to have been caused by an act or directive of the owner or owner's representative. The owner denies that a change in the scope of work has occurred and contests the contractor's entitlement to additional compensation.

contingency allowance   A sum or percentage included in the project budget designated to cover unpredictable or unforeseen items of work, or changes in the work subsequently required by the owner.

contingent liability   (1) Liability that is not absolute and fixed, but is dependent upon the occurrence of some uncertain future event or the existence of an uncertain specified condition. (2) Particularly in insurance law, liability imposed upon an individual or entity because of injuries or damages caused by persons or entities, other than employees, for whose acts or omissions the first party may be held legally responsible.

contract   A legally enforceable promise or agreement between two or among several persons.

contract documents   The owner-contractor agreement, the conditions of the contract general, supplementary and other conditions, the drawings, the specifications, and all addenda issued prior to and all modifications issued after execution of the contract; and any other items that may be specifically stipulated as being included in the contract documents.

contract limit   A limit line or perimeter line established on the drawings or elsewhere in the contract documents for substantial completion of the work, including authorized adjustments thereto. The contract time may be adjusted only by change order.

contractor   (1) One who contracts. (2) In construction terminology, the person or entity responsible for performing the work and identified as such in the owner/ contractor agreement.

contractor's equipment floater   Insurance which provides protection to the contractor against loss or damage to equipment because of fire, lightening, tornado, flood, collapse of bridges, perils of transportation, collision, theft, landslide, overturning, riot, strike, and civil commotion.

contractor's liability insurance   Insurance purchased and maintained by the contractor to protect the contractor from specified claims which may arise out of or result from the contractor's operations under the contract, whether such operations are by the contractor or by any subcontractor or by anyone directly or indirectly employed by any of them, or by anyone for whose acts any of them may be liable.

contractor's protective liability insurance   A contingent insurance that protects a contractor against claims resulting from accidents caused by subcontractors or their employees, for which the contractor may be held liable.

cost breakdown   A schedule prepared by the contractor showing how the contract sum is divided among the various divisions of the work, which is filed before the work is started. The architect will use this as a basis for approving applications for payment and waivers. It should be accurate and kept up to date, reflecting any changes in the contract.

cost code   A systematic classification and categorization of all items of work or cost pertaining to a project.

cost report A report detailing the costs of the project into its various codes and projecting the final cost of each activity to indicate the probable cost of the project upon completion.

"cost" rule The cost rule is the measure of damages for the breach of a building construction contract. It is ordinarily such sum as is required to make the building conform to the contract.

coverage Dependable estimates or firm bids for portions of a construction project.

cut The act of reducing a price, bid, or estimate.

daily log The daily record of the happenings on a construction project. The items typically recorded include weather conditions, names of subcontractors on the site, number of people working for each subcontractor, visitors, problems encountered, work performed, start time, finish time, and verbal directions given by the owner or owner's representative.

delay A slip in the approved or mandated construction schedule.

deductible When applied to insurance, the amount of money not covered by the insurance company.

deductive alternate An alternate bid resulting in a deduction from the same bidder's base bid.

deductive change A directed change whereby the owner deletes work from the original scope of work established in the contract documents.

design/build A method of organizing a construction project where the owner awards a single contract to one firm that designs and builds the facility. After conceptual design documents and a preliminary cost estimate have been prepared, the parties negotiate a guaranteed maximum price for a cost-plus-fixed-fee contract. At this stage, the project is defined only in terms of design parameters and standards. As work progresses, the owner pays all allowable costs up to the guaranteed maximum price. Design/build techniques are used primarily on heavy, engineered construction projects such as industrial facilities and utility plants.

differing site conditions clause A clause in a construction contract authorizing the contractor to receive an equitable price adjustment if the contractor encounters differing site conditions. In the absence of such a clause, the contractor must complete the work in accordance with the contract documents at the contract price, regardless of the physical conditions encountered in the field. A material misrepresentation of physical conditions may be a breach of contract by the owner, however.

differing site conditions type 1 Actual site conditions that differ materially from conditions affirmatively represented in the contract documents. The name is derived from the differing site conditions clause found in federal government construction contracts.

differing site conditions type 2 Actual physical conditions that differ materially from the conditions that a prudent, experienced contractor would reasonably expect to encounter on a project of that nature.

directed change A change ordered by the project owner and acknowledged by the owner to be a change in the scope of work.

disruption An act, failure to act, or directive from the owner that forces the contractor to perform work in a sequence that varies from the conventional or logical sequence in which the contractor had planned to perform the work. Generally, the contractor is entitled to recover the increased costs resulting from the disruption.

division of work The prime contractor's allocation of work among the various subcontractors. The prime contractor has sole responsibility for dividing the work among subcontractors in such a way that there is no duplication and there are no gaps. The project owner and its architect or engineer have no obligation to prepare plans and specifications in a way that facilitates the award of subcontracts.

drawings Graphic and pictorial documents showing the design, location, and dimensions of the elements of a project. Drawings generally include plans, elevations, sections, details, schedules, and diagrams. When capitalized, the term refers to the graphic and pictorial portions of the contract documents.

due care A legal term indicating the requirement for a professional to exercise reasonable care, skill, ability, and judgment under the circumstances. Performance of duties and services must be consistent with the level of reasonable care, skill, ability, and judgment provided by reputable professionals in the some geographical area at the same period of time. Failure to exercise due care constitutes negligence.

early finish Early start plus activity duration (earliest date an activity can finish).

early start Early event time of preceding event (earliest time an activity can start).

economic loss doctrine This provides that there is no recovery of economic damages in tort where there is no property damage or personal injury.

element The smallest division of a breakdown.

elevation (1) A two-dimensional graphic representation of the design, location, and certain dimensions of the project, or parts thereof, seen in a vertical plane viewed from a given direction. (2) Distance above or below a prescribed datum of reference point.

employers liability insurance Insurance protection for the employer against claims by employees or employees' dependents or damage which arises out of injuries or diseases sustained in the course of their work, and which are based on common law negligence rather than on liability under workers' compensation acts.

equitable estoppel If a contractor reasonably relies on a quotation submitted by a subcontractor and incorporates the quoted price into a firm bid, the doctrine of "equitable estoppel" will prevent the subcontractor from reneging on its quotation if the contractor's bid is accepted by the project owner and the contractor is called upon to perform the work at its bid price.

equitable price adjustment A general term, frequently used in construction contract changes clauses, referring to the price increase (or decrease, in the event of a deductive change) that will leave the contractor in the same financial position it would have been in but for the change in the scope of work.

escalation The increase in the contractor's cost of labor and materials that occurs during a period of delay. If the delay is compensable, this price escalation will be a major component of the contractor's recoverable delay damages.

estimate A time/cost prediction; the act of preparing an estimate; the estimate itself.

estimator The person responsible for all activities involved in preparing the contractor's bid.

excusable delay A delay in the contractor's performance of the work which results from a factor that is beyond the control and without the fault of either owner or contractor. Unforeseeably severe weather is the classic example of excusable delay. If an excusable delay occurs, the contractor is entitled to an extension of the performance period for a commensurate amount of time but is not entitled to any additional compensation.

experience modification factor The multiplier relating to the safety record of an employer that is applied to the standard workers' compensation rate to determine the actual premium owed by the contractor.

extended coverage insurance An endorsement to property insurance policy which extends the perils covered to include windstorm, hail, riot, civil commotion, explosion (except steam boiler), aircraft, vehicles, and smoke.

extension The product of a quantity multiplied by a unit price; the completion of a mathematical equation.

extras Construction costs outside of contract sum. A contractor should not proceed with any extras until a change order is issued by the architect.

fidelity bond A surety bond that reimburses an employer named in the bond for loss sustained by reason of the dishonest acts of an employee covered by the bond.

field Occupation such as a trade, profession, or specialty; the actual construction site.

filed subbids A system used on some public projects whereby trade contractors submit bids to the project owner for specified portions of the project. The general contractor must use the low subbidders when preparing its bid and performing the work unless the general contractor intends to perform that portion of the work with its own forces.

final acceptance The owner's acceptance of the project from the contractor upon certification by the architect of final completion. Final acceptance is confirmed by the making of final payment unless otherwise stipulated at the time of making such payment.

final completion Term denoting that the work has been completed in accordance with the terms and conditions of the contract documents.

final inspection Final review of the project by the architect to determine final completion, prior to issuance of the final certificate for payment.

final payment Payment made by the owner to the contractor, upon issuance by the architect of the final certificate for payment, of the entire unpaid balance of the contract sum as adjusted by change orders. See also **final acceptance**.

flow-down clause A clause included in some subcontracts which states that the subcontractor assumes toward the prime contractor all the duties and obligations the prime contractor has assumed toward the project owner under the terms of the prime contract. Generally, this is interpreted to include notice requirements and other procedural aspects of the prime contract. On federal construction projects, however, it is interpreted to apply only to the definition of the contraction work itself.

force account Work performed by a contractor utilizing his own labor force and equipment.

forward pricing The assignment of a contract price adjustment to a change order before the actual costs of the change are known. Forward pricing is based on estimates. The owner and contractor agree on the price adjustment before the changed work is actually performed.

fringe benefits An employment benefit given in addition to regular wages or salary. These benefits may be paid directly to the employee or may be paid by the employer to various agencies on behalf of the employees.

general conditions These define the rights, responsibilities, and relations of all parties to the contraction contract.

general conditions expense All costs that can be readily charged to a specific project.

general contractor The contractor who has a direct or prime contract with the project owner. Responsible for completing the entire project in strict accordance with the plans, specifications, terms, and conditions of the construction contract.

guarantee A contractual obligation to replace or repair defective items for a stated period of time. Guarantees usually apply to specific pieces of equipment or specific portions of the work, such as the roof. Guarantees are frequently backed up by bonds. Guarantees are commonly furnished by equipment or material suppliers to the prime contractor, who in turn assigns the guarantee to the project owner upon completion of the work.

inconsistent dispute provisions A problem that arises when the various contracts on a construction project call for different methods of resolving disputes. This makes it impossible to get all parties into the same forum at the same time, even though all the disputes are based on the same factual occurrences. It also increases the likelihood of inconsistent or unjust results.

indemnification A contractual obligation by which one person or entity agrees to secure another against loss or damage from specified liabilities.

indemnification implied An indemnification that is implied by law rather than arising out of a contract to provide indemnification.

inexcusable delay A delay which is caused by an event within the control of the contractor and beyond the control of the owner. The contractor gets neither time nor money and the owner may recover damages.

installation floater policy Provides protection to the contractor against loss resulting from the collapse of a structure during erection.

instructions to bidders The document that states the procedures to be followed by all bidders.

invitation-to-bid The formal documents issued by the owner stating that the owner is inviting contractors to bid on a project. The invitation-to-bid usually includes the scope of work; date and time for bid; prequalification requirements; amount of deposit for plans and specifications; date and time of the prebid conference; and any other information that the owner feels is pertinent.

item A subdivision of the breakdown, smaller than a category, but larger than an element.

job A construction project.

job site Within the property lines of a construction project.

labor Any work performed by an employee.

labor and material payment bond A bond of the contractor in which a surety guarantees to the owner that the contractor will pay for labor and materials used in the performance of the contract. The claimants under the bond are defined as those having direct contracts with the contractor or and subcontractor. A labor and material payment bond is sometimes referred to as a payment bond.

late finish Late event time of following event. (Latest time an activity can finish with the project staying on schedule.)

late start Late finish minus activity duration. (Latest time an activity can start and stay on schedule.)

left on the table The dollar difference between the low bid and the next bid above.

legitimate Ethical and legal.

let or sublet   Issue a contract for a portion of the project.

letter of intent   Written prior to drawing up complete contract documents to tell the contractor a formal agreement is forthcoming. Often this authorizes certain preliminary work such as filing for building permits and site clearance.

liability insurance   Insurance that protects the insured against liability on account of injury to the person or property or another.

licensed contractor   A person or organization certified by governmental authority, where required by law, to engage in construction contracting.

licensing requirements   The licensing required by each individual state for individuals or organizations performing contraction work within that state. Many states require separate licenses for the various trades or for general contractors undertaking very large projects. The possession of a valid state license is usually a legal precondition to a contractor's recovery of payment under the contract. Some states strictly enforce this requirement. Others accept a contractor's good faith effort at substantial compliance.

lien   A right created by state statute that allows parties furnishing labor or materials to a project to obtain a security interest in the real estate in order to protect their entitlement to payment. Liens are generally available to prime contractors, subcontractors, and material suppliers. Unpaid parties may perfect liens only by fully and carefully complying with the requirements of the applicable state statute.

lien waiver   A document whereby the contractor acknowledges it has been paid for completed work and waives its statutory right to assert a mechanic's lien against the owner's project. Generally, a contractor is required to furnish partial lien waivers upon receipt of progress payments and a final lien waiver upon final payment. Contractors are also usually required to furnish the owner with lien waivers from all the subcontractors and suppliers with whom the contractor has done business.

line item   Any item on a price-out sheet and all of the quantities, unit prices, and extensions.

liquidated damages   A sum agreed to by owner and contractor, chargeable to the contractor, for damages suffered by the owner because of contractor's failure to fulfill his contract obligations.

lis pendens   A notice of pending litigation.

litigation   The resolution of disputes through the use of the judicial system. State and federal statures define the jurisdiction of the various courts, determining where lawsuits may be initiated.

loss of use insurance   Insurance protecting against financial loss during the time required to repair of replace property damaged or destroyed by an insured peril.

low bid   Bid stating the lowest bid price, including selected alternates, and complying with all bidding requirements.

lowest responsible bidder   The bidder who submits the lowest bona fide bid and is considered by the owner and the architect to be fully responsible and qualified to perform the work for which the bid is submitted.

lowest responsive bid   The lowest bid that is responsive to and complies with the bidding requirements.

lump sum   An item or category priced as a whole rather than broken down into its elements.

lump sum bid A bid priced as a whole rather than broken down into its elements.

maintenance bond Guarantees owner that contractor will rectify defects in workmanship or materials reported to him within a specified period following final acceptance of the work under contract. Often this is included as part of the performance bond and usually runs for one year. May also be issued for longer periods covering specific phases of the work, such as roofing, curtain wall, paving, etc.

mark-up A contractor's overhead and profit.

mechanic's lien A lien on real property created by statute in all states in favor of person supplying labor or materials for a building or structure for the value of labor or materials supplied by them. In some jurisdictions a mechanic's lien also exists for the value of professional services. Clear title to the property cannot be obtained until the claim for the labor, materials, or professional services is settled.

med-arb Employing aspects of both mediation and arbitration, the parties mediate the dispute with an agreement to submit any unresolved issues to binding arbitration later.

mediation An attempt by an objective third party to negotiate or facilitate a resolution of a dispute. The recommendation of the mediator is not binding on the parties.

Miller Act A federal statute requiring prime contractors on federal construction projects in excess of $25,000 to furnish payment bonds protecting subcontractors and suppliers. The Miller Act served as the prototype for most state public works payment bond statutes.

minitrial A structured process in which management or executive level representatives of the disputants attend a relatively short and informal hearing, usually assisted by a neutral advisor.

minor changes in the work Changes of a minor nature in the work not involving an adjustment in the contract sum or an extension of the contract time and not inconsistent with the intent of the contract documents.

mistaken bid A bid that contains an error of a clerical or mathematical, as opposed to judgmental, nature. If the bidder can demonstrate the existence of the mistake and show that the mistake occurred despite its own reasonable precautions, the bidder will be allowed to withdraw the bid and the bid bond will not be in jeopardy.

modification (to the contract documents) (1) A written amendment to the contract signed by both parties. (2) A change order. (3) A written or graphic interpretation issued by the architect. (4) A written order for a minor change in the work issued by the architect.

negligence Failure to exercise due care under the circumstances. Legal liability for the consequences of an act or omission frequently depends upon whether or not there has been negligence. See also **due care**.

no-damage-for-delay clause A contractual clause whereby the owner disclaims any liability for the contractor's damages or increased performance consists resulting from delay of any nature, including delay caused by the owner's breach of contract. With certain exceptions, no-damage-for-delay clauses are enforced against contractors.

nonconforming work Work that does not fulfill the requirements of the contract documents.

nonexcusable delay A delay which results from the contractor's failure to properly carry out its obligations under the contract. The contractor receives no additional compensation and no extension of the performance period.

notice to bidders A notice contained in the bidding requirements informing prospective bidders of the opportunity to submit bids on a project and setting forth the procedures for doing so.

notice to proceed Written communication issued by the owner to the contractor authorizing him to proceed with the work and establishing the date of commencement of the work.

offer Promise to do something in exchange for a price.

owner's representative The person designated as the official representative of the owner in connection with a project.

overhead (contractor) Project costs for all items that cannot be classified as "direct field costs". These costs include, but are not limited to, home office physical plant services, support and *pro rata* salaries, project management, clerical, accounting, general liability and comprehensive insurance costs, and outside services.

overrun The amount by which an item costs more than estimated; the amount a quantity increases over the estimated quantity.

owner The person or institution who initiates the project, furnishes the concept, the land, and the funds, and holds title of ownership.

parole evidence rule The doctrine provides that where the parties reduced their agreement to a written contract, all prior negotiations and understandings (whether written or oral) were merged into the contract and thus could not be used to supplement, vary, or contradict the terms of the contract.

partial occupancy Occupancy by the owner of a portion of a project prior to final completion. See final completion.

partial payment See progress payment.

partnering A process which attempts to establish working relationships among the parties in a construction project through a mutually developed, formal strategy of commitment and communication. It attempts to create an environment where trust and teamwork prevent disputes; foster a cooperative bond to everyone's benefit; and enable the project to be completed successfully.

payment and performance bond A bond of the contractor in which a surety guarantees to the owner that the work will be performed in accordance with the contract documents. Except where prohibited by statute, the performance bond is frequently combined with the labor and material payment bond.

penal sum The amount named in contract or bond as the penalty to be paid by a signatory thereto in the event he fails to perform his contractual obligations.

performance bond Guarantees the owner that, within limits, the contractor will perform all work in accordance with the contract documents and that the owner will receive the project built in substantial agreement with the documents.

plans Drawings in plan view. A portion of a total set of drawings.

plug-in A temporary figure in an estimate price-out sheet to be used until a more dependable one is obtained.

prebid conference A meeting held prior to bid opening for the purpose of answering questions that bidders may have with respect to the bid documents.

prebid site inspection The contractor's obligation, established in the bid documents or other contract documents, to thoroughly inspect the job site prior to submitting a bid. Contractors are held responsible for any site information that could have been determined during a reasonable site inspection, regardless of whether the contractor actually conducted such an inspection. Contractors are not expected to determine the nature of hidden, or latent, conditions and are not expected to conduct independent tests.

preconstruction conference (precon) Meeting held with the owner, architect, and contractor in attendance prior to the start of construction to discuss schedule, progress payments, submittals, staging, etc.

prequalification of prospective bidders The process of investigating the qualifications of prospective bidders on the basis of their competence, integrity, and responsibility relative to the contemplated project.

price out The activity of applying dollar values to the items in a take-off; the final estimate sheet showing all the dollar values.

prime bid A bid for the total project; general contractor's bid rather than a subcontractor's bid.

prime contractor See **general contractor**.

privity of contract A direct contractual relationship. There used to be a legal requirement, now virtually obsolete, that a party be in privity of contract with another party before the party could be held liable to that other party.

procure To obtain or receive, as a construction contract.

production Quantity of work performed, usually related to a time period, such as square feet per hour (sq ft/hr).

production rates The number of units of work produced by a unit of equipment or a person in a specified unit of time.

profit The amount of money, if any, that a contractor retains after he has completed a project and has paid all costs for materials, equipment, labor, overhead, taxes, insurance, etc.

progress payment Interim payments made to the contractor reflecting the value of work completed to date. The value is usually determined by the "percentage of completion" method but may be measured according to contractually stipulated unit prices or a schedule of values.

progress schedule A diagram, graph, or other pictorial or written schedule showing proposed and actual dates of starting and completion of the various elements of the work.

project (1) The total construction of which the work performed under the contract documents may be the whole or a part. (2) The total furniture, furnishings, and equipment and interior construction of which the work performed under the contract documents may be the whole or a part.

project application for payment Certified requests for payment from individual contractors on a construction management project, assembled for certification by the architect, and submitted to the owner.

project cost Total cost of the project including construction cost, professional compensation, land costs, furnishings and equipment, financing, and other charges.

project manager The person responsible for managing the project's scope, budget, schedule, quality, and safety aspects.

project manual The volume(s) of documents prepared by the architect for a project which may include the bidding requirements, sample forms and conditions of the contract and the specifications.

project superintendent See **superintendent**.

promissory estoppel An equitable doctrine that holds that if one party changes its economic position in reasonable reliance on the promise or affirmative representation of another party, the second party is precluded from reneging on its promise. In construction contracting, this doctrine is applied to enable prime contractors to hold subcontractors to their oral, prebid quotations, even though no subcontract agreement has yet been formed.

property insurance Coverage for loss or damage to the work at the site caused by the perils of fire, lighting, extended coverage perils, vandalism, and malicious mischief and additional perils (as otherwise provided or requested). Property insurance may be written at the start of a project in a predetermined amount representing the insurable value of the work (consisting of the contract sum less the cost of specified exclusions) and adjusted to the final insurable cost on completion of the work, or (2) the reporting form in which the property values fluctuate during the policy term, requiring monthly statements showing the increase in value of work in place over the previous month. See also **extended coverage insurance; special hazards insurance.**

public liability insurance Insurance covering liability of the insured for negligent acts resulting in bodily injury, disease, or death of persons other than employees of the insured, and/or property damage. See also **comprehensive general liability insurance; contractor's liability insurance.**

punch list The list of items that remains to be completed or corrected after the contractor has achieved substantial completion.

quality assurance The application of standards and procedures to ensure that a product or a facility meets or exceeds desired performance criteria.

quality control This process includes (1) setting specific standards for construction performance, usually through the plans and specifications; (2) measuring variances from the standards; (3) taking action to correct or minimize adverse variances; and (4) planning for improvements in the standards themselves and in conformance with the standards. In other words, once the architects and engineers have set the criteria for construction, quality control ensures that the physical work conforms to those standards.

quantity survey The activity of determining quantities from drawings and specifications.

quantity variations A variation, usually on a unit-price contract, between the estimated quantities of work stated in the contract and the actual quantities encountered in the field. A significant underrun can pose a financial hardship for the contractor, whereas an overrun can produce a financial windfall. Consequently, many unit-price contracts call for a price adjustment if the actual quantity varies from the estimated quantity by more than a stated percentage amount.

quote To make an offer at a guaranteed price.

quotation A price quoted by a contractor, subcontractor, material supplier, or vendor to furnish materials, labor, or both.

record drawings Construction drawings revised to show significant changes made during the construction process, usually based on marked-up prints, drawings and other data furnished by the contracts to the architect. Preferable to as-built drawings.

rejection of work (by the architect): the act of rejecting work that is defective or does not conform to the requirements of the contract documents.

release of lien Instrument executed by a person or entity supplying labor, materials, or professional services on a project that releases that person's or entity's mechanic's lien against the project property. See also **mechanic's lien; waiver of lien**.

rendering A drawing of a project or portion thereof with an artistic delineation of materials, shades, and shadows.

request for information A formal request from the contractor to the owner requesting additional information or clarification of intent regarding the project.

retainage A contractually stipulated amount, usually 5 or 10%, that is withheld or retained by the project owner from each progress payment. The purpose of retainage is to provide the owner with security in the event of defective or incomplete work by the contractor. Retainage is released to the contractor upon final acceptance of the completed project.

schedule of values A statement furnished by the contractor to the architect reflecting the portions of the contract sum allocated to the various portions of the work and used as the basis for reviewing the contractor's applications for payment. See also **cost breakdown**.

scope of work The project, drawings, and specifications are identified; the architect is listed. The contractor agrees to furnish all material and perform all of the work for the project in accordance with the contract documents.

section (of drawings) A drawing of a surface revealed by an imaginary plane cut through the project, or portion thereof, in such a manner as to show the composition of the surface as it would appear if the part intervening between the cut plane and the eye of the observer were removed.

section (of specifications) A subdivision of a division of the specification that should cover the work of no more than one trade.

separate contract A contract let by the owner directly to a contractor other than the general contractor. Usually for special equipment, landscaping, interior furnishings, etc.

separate contractor A contractor on a project having a separate contract with the owner.

sequence of work The prime contractor's scheduling of subcontractor's work in such a way that the project is constructed in a logical, orderly fashion and no subcontractor is delayed or disrupted by the untimely performance of another subcontractor. The prime contractor has sole responsibility for scheduling and coordinating the work of its subcontractors.

sequential delay A delay that recognizes an overall delay as caused by different types of delaying events operating sequentially rather than concurrently and to which the rules of each type of delay will apply to each sequential delaying effect.

shop drawings Drawings prepared at the contractor's expense showing how specific items shall be fabricated and/or installed. These usually are prepared by the subcontractors and are standard for such items as steel, reinforcing, miscellaneous and ornamental iron, cut stone, millwork, partitions, door frames, etc. They must be submitted to the architect for his approval.

site Geographical location of the project, usually defined by legal boundaries.

site condition A physical condition, existing at the site where the work is to be performed, that may affect the performance of the work. Site conditions include subsurface soil and water conditions, existing utilities, and conditions in existing structures.

special conditions A section of the conditions of the contract, other than general conditions and supplementary conditions, that may be prepared to describe conditions unique to a particular project.

special hazards insurance Insurance coverage for damage caused by additional perils or risks to be included in the property insurance at the request of the contractor or at the option of the owner. Examples often included are sprinkler leakage, collapse, water damage, and coverage for materials in transit to the site or stored off the site. See **property insurance**.

specifications A part of the contract documents contained in the project manual consisting of written requirements for materials, equipment, construction systems, standards, and workmanship.

spreadsheet Large, wide sheet with many columns that is used to tabulate all estimates and sub-bids when putting a bid together.

staging The process of storing and moving materials on the job site from the truck to final installation.

statute of limitations A stature specifying the period of time within which legal action must be brought for alleged damage or injury, or other legal relief. The lengths of the periods vary from state to state and depend upon the type of legal action. Ordinarily, the period commences with the occurrence of the damage or injury, or discovery of the act resulting in the alleged damage or injury. In construction cases, many jurisdictions define the period as commencing with completion of the work or services performed in connection therewith.

statutory bond A bond, the form or content of which is prescribed by statute.

step negotiation A form of informal dispute resolution where negotiation between the party representatives with primary responsibility for resolving the dispute in question is unsuccessful. The dispute is referred progressively upward for negotiation between representatives at the next highest level in the management chain.

sub-bid A bid offered by a subcontractor.

sub-subcontractor A person or entity who has a direct or indirect contract with a subcontractor to perform any of the work at the site.

subcontract Agreement between a prime contractor and a subcontractor for a portion of the work at the site.

subcontractor A person or entity who has a direct contract with the contractor to perform any of the work at the site.

subcontractor bonds A performance bond given by a subcontractor to the general contractor guaranteeing performance of contract and payment of labor and material bills.

submit the bid Deliver the bid to a sponsor.

subrogation The substitution of one person or entity who has a direct contract with the contractor to perform any of the work at the site.

The substitution of one person for another with respect to legal rights such as a right of recovery. Subrogation occurs when a third person, such as an insurance company, has paid a debt of another or claim against another and succeeds to all legal rights that the debtor or person against whom the claim was asserted may have against other persons.

substantial completion That time when the project is sufficiently complete to enable the owner to occupy the project and make beneficial use of the project for its intended purpose.

substitution An equivalent material, product, or item of equipment offered in lieu of that specified.

subsurface investigation (Sometimes called geotechnical investigation). The soil boring and sampling program, together with the associated laboratory tests, necessary to establish subsurface profiles and the relative strengths, compressibility, and other characteristics of the various strata encountered within depths likely to have an influence on the design of the project.

successful bid A bid that is accepted by a sponsor for award of contract; a low bid (assuming that there will be an award of contract).

superintendence The work of the contractor's representative, not to be confused with that of the architect (administration) or clerk of the works.

superintendent Contractor's representative at the site who is responsible for continuous field supervision, coordination, completion of the work and, unless another person is designated in writing by the contractor to the owner and the architect, the prevention of accidents.

supervision Direction of the work by contractor's personnel. Supervision is neither a duty nor a responsibility of the architect as part of professional services.

supplementary general conditions Since conditions vary by locality, these are used to amend or supplement portions of the general conditions.

supplier A person or organization who supplies materials or equipment for the work, including that fabricated to a special design, but who does not perform labor at the site.

supply bond A guarantee to the owner by a manufacturer that the materials delivered comply with the contract documents.

surety A person or organization who, for a consideration, promises in writing to make good the debt or default of another.

surety bond A legal instrument under which one party agrees to answer to another party for the debt, default, or failure to perform of a third party.

suspension An owner's directive to the contractor to suspend operations. The owner's right to suspend work is usually established in the construction contract itself. Frequently, the owner is not required to pay additional compensation if the suspension is only for a "reasonable" period of time. The reasonableness of the duration of a suspension is a case-by-case factual determination.

take-off The activity of determining quantities from drawings and specifications; the actual quantity lists.

tender A formal offer of a bid.

termination for convenience A project owner's right, established in the terms of the contract, to remove the contractor from the project despite the absence of any contractor wrongdoing. Termination for convenience clauses are intended to give the owner the flexibility to abort a project for reasons of its own convenience. In the event of a termination for convenience, the owner must pay the contractor for all work completed prior to the effective date of termination, as well as other contractually stipulated termination costs. The contractor has no right to sue for breach of contract or to recover lost profit on work it was unable to perform.

termination for default A project owner's right, established in the terms of the contract, to remove the contractor from the project because of the contractor's material breach of the contract. In the event of a termination for default, the owner has no further contractual obligations to the contractor. The contractor, however, is obligated to reimburse the owner for costs incurred in completing the work to the extent such costs exceed the original contract price.

time (as the essence of the construction contract) Time limits or periods stated in the contract. A provision in a construction contract that "Time is of the essence of the contract" signifies that the parties consider that punctual performance within the time limits or periods in the contract is a vital part of the performance and that failure to perform on time is a breach for which the injured party is entitled to damages in the amount of the loss sustained.

time of completion Date established in the contract, by a calendar date or by number of dates, for substantial completion of work. See also **substantial completion**.

timely completion Completion of the work or designated portion thereof on or before the date required.

total float Late finish minus early finish.

trade A subdivision of a breakdown, composed of workmen specializing in a particular skill.

umbrella liability insurance Insurance providing excess liability coverage over existing liability policies, such as employer's liability, general liability, or automobile liability, and providing direct coverage for many losses uninsured under the existing policies after a specified deductible is exceeded.

unabsorbed home office overhead The portion of the contractor's daily fixed office overhead costs that would have been absorbed by billings to a particular contract but for an owner-caused delay in the contractor's performance of that contract. Unabsorbed home office overhead is a major component of a contractor's delay damages in a compensable delay situation.

unbalanced bid A bid is unbalanced if unit prices or components of a lump sum price do not carry an appropriate share of the total direct cost, overhead, and profit. A bid is materially unbalanced if acceptance of that bid creates a risk that the project owner will not obtain the lowest possible price. Public project owners are not obligated to accept a low bid if that bid is materially unbalanced.

unconditional lien waiver Lien waiver that is valid with no conditions necessary.

Uniform Commercial Code A model statute that has been adopted in virtually every state and covers the purchase and sale of goods. It does not apply to construction contracts but does apply to agreements between contractors and their equipment or material suppliers. The Uniform Commercial Code addresses vital matters such as offer and acceptance, implied warranties, and the disclaimer of warranties.

uniform construction index   A published system for coordination of specification sections, filing of technical data and product literature, and construction cost accounting organized into sixteen divisions.

unit prices   Amounts stated in a contract as prices for materials per unit of measurement or services as described in the contract documents.

upset price   An amount agreed to by the contractor as a maximum cost to perform a specific project. Used on a cost-plus-a-fee jobs.

"value" rule   The value rule is generally applied as an exception to the "cost" rule. In situations where the contractor is successful in proving that its performance of the contract constituted substantial performance, i.e., that the extent of deviation from the contract was minor.

vandalism and malicious mischief insurance   Insurance against loss or damage to the insured's property caused by willful and malicious damage or destruction.

vendor   A person or organization who furnishes materials or equipment not fabricated to a special design for the work.

waiver   The contractor's waiver of the right to receive additional compensation under the differing site conditions clause. The most common form of waiver is the contractor's failure to give the owner prompt notice of a differing site condition prior to disturbing that condition.

waiver of lien   An instrument by which a person or organization who has or may have a right of mechanic's lien against the property of another relinquishes such right. See also **mechanic's lien; release of lien**.

warranty   A contractual commitment to replace or repair any defective work. All construction contracts contain an implied (or unstated) warranty that the workmanship will meet the accepted standards of the construction industry.

work   The completed construction required by the contract documents, including all labor necessary to produce such construction, and all materials and equipment incorporated or to be incorporated in such construction.

working drawings   Those actual plans (drawings and illustrations) from which the project is to be built. They contain the dimensions and locations of building elements and materials required and delineate how they fit together.

workers' compensation insurance   Insurance covering the liability of an employer to employees for compensation and other benefits required by workers' compensation laws with respect to injury, sickness, disease, or death arising from their employment.

XCU (Insurance terminology)   Letters that refer to exclusions from coverage for property damage liability arising out of (1) explosion or blasting, (2) collapse of or structural damage to any building or structure, and (3) underground damage caused by and occurring during the use of mechanical equipment.

Bibliography

Ahuga, Hira N., and Walter J. Campbell, *Estimating: From Concept to Completion*, Prentice-Hall, Englewood Cliffs, New Jersey, 1988.

Atcheson, Daniel, *Estimating Earthwork Quantities*, Norseman Publishing Company, Lubbock, Texas, 1986.

Avery, Craig, (ed.), *Concrete Construction & Estimating*, Craftsman Book Company, Carlsbad, California, 1975.

Barrie, Donald S., and Boyd C. Paulson, *Professional Construction Management*, 2nd ed., McGraw-Hill, New York, 1984.

Bernold, Leonard E., and John F. Tresular "Vendor Analysis for Best Buy in Construction," *ASCE Journal of Construction Engineering and Management*, (Vol. 117, No. 4), pp. 645-658, December, 1991.

Bourges, G. Patrick, *et al.*, *Walker's Quantity Surveying and Basic Construction Estimating*, Frank R. Walker Company, Chicago, Illinois, 1981.

Carr, Robert, "Cost Estimating Principles," *ASCE Journal of Construction Engineering and Management*, (Vol. 115, No. 4), pp. 545,-551 December, 1989.

Chick, David, "The Changing Role of the Estimator," *Cost Engineering*, (Vol. 34, No. 7), July, 1992.

Clark, John E., *Structural Concrete Cost Estimating*, McGraw-Hill Book Company, New York, 1983.

Cleveland, Allan B., (ed.), *Means' Man-hour Standards*, R.S. Means Company, Inc., Kingston, Massachusetts, 1983.

Clough, Richard H., *Construction Contracting*, 2nd ed., Wiley - Interscience, New York, 1969.

Construction Dictionary, 6th ed., Greater Phoenix, Arizona Chapter, National Association of Women in Construction, Phoenix, Arizona, 1985.

Construction Industry Cost Effectiveness Report, (Vol. 23), The Business Roundtable, New York, 1980-1982.

"Contractor Survey Finds That Specs Don't Measure up: Language Barriers, Shrinking Fees, and Pace of Change Cited as Reasons for the Problems," *Engineering - News Record*, June 17, 1991.

Cook, Paul J., *Bidding for the General Contractor*, R.S. Means Company, Inc., Kingston, Massachusetts, 1985.

------, *Estimating for the General Contractor*, R.S. Means Company, Inc., Kingston, Massachusetts, 1985.

Cox, Billy J., and F. William Horsley, *Means Square Foot Estimating*, R.S. Means Company, Kingston, Massachusetts, 1983.

Crespin, Vick S., C. Dawson Zeigler, Brisbane H. Brown, Jr., and Luther J. Strange, *Walker's Manual for Construction Cost Estimating*, Frank R. Walker Company, Lisle, Illinois, 1989.

Dagostino, Frank R., *Estimating in Building Construction*, 2nd ed., Reston publishing Company, Reston, Virginia, 1978.

de Neufville, Richard, and Daniel King, "Risk and Need for Work Premiums in Contractor Bidding", *ASCE Journal of Construction Engineering and Management*, (Vol. 117, No. 4), pp. 659-673, December, 1991.

Dunham, Clarance W., Robert D. Young, and Joseph T. Rockrath, *Contracts, Specifications and Law for Engineers*, 3rd ed., McGraw-Hill, New York, 1979.

Erickson, Carl A., and LeRoy T. Roger, "Estimating - State-of-the-Art," *American Society of Civil Engineers Journal of the Construction Division*, September, 1976, pp. 455-463.

Farid, Foad, and L.T. Royer, "Fair and Reasonable Mark-up (FARM) Pricing Model", *American Society of Civil Engineers Journal of Construction Engineering and Management*, (Vol. III, No. 4), pp. 374-390, December, 1985.

Fee, Sylvia Hollman, *Means Landscape Estimating*, R.S. Means Company, Kingston, Massachusetts, 1987.

Fico, Karen, (ed.), Business Failure Record - 1990, The Dunn & Bradstreet Corporation, New York, 1990.

------, Business Failure Record - 1991, The Dunn & Bradstreet Corporation, New York, 1991.

------, Business Failure Record - 1992, The Dunn & Bradstreet Corporation, New York, 1992.

Foster, Norman, and Trauner, Theodore J., and Vespe, Rocco, R., *Construction Estimates from Take-off to Bid*, 4th ed., McGraw-Hill Book Company, New York, 1995.

Frein, Joseph P., *Handbook of Construction Management and Organization*, 2nd ed., Van Nostrand Reinhold Company, New York, 1980.

Gladstone, John, <u>Mechanical Estimating Guidebook</u>, 4th ed., McGraw Hill Book Company, New York, 1970.

<u>Glossary of Construction Terms</u>, 1992 Edition, Association of Construction Inspectors, San Francisco, California, 1992.

Goldman, Jeffrey M., (ed.), <u>Means Estimating Handbook</u>, R.S. Means Company, Inc., Kingston, Massachusetts, 1990.

Griffis, F.H., "Bidding Strategy: Winning over Key Competitors," <u>ASCE Journal of Construction Engineering and Management</u>, (Vol. 118, No. 1), pp. 151-165, March, 1992.

Grow, Thomas A., <u>Construction: A Guide for the Profession</u>, Prentice Hall, Englewood Cliffs, New Jersey, 1975.

Jackson, Patricia, (Editor-in-Chief), <u>Means Assemblies Cost Data - 18th Annual Edition</u>, R.S. Means Company, Kingston, Massachusetts, 1993.

Jervis, Bruce M., and Paul Levin, <u>Construction Law - Principles and Practice</u>, McGraw-Hill Book Company, New York, 1988.

Kangari, Roozbeh, "Business Failure in the Construction Industry," <u>ASCE Journal of Construction Engineering and Management</u>, (Vol. 114, No. 2), pp. 172-190, June, 1988.

Kolkoski, Rynold, <u>Masonry Estimating</u>, Craftsman Book Company, Carlsbad, California, 1988.

<u>Labor Estimating Manual, 1992 Volume</u>, Mechanical Contractor's Association of America, Inc., Rockville, Maryland, 1992.

Levy, Sidney M., <u>Project Management in Construction</u>, McGraw Hill Book Company, New York, 1987.

Lew, Alan E., <u>Means Interior Estimating</u>, R.S. Means Company, Kingston, Massachusetts, 1987.

Lifshitz, Judah, and Jean Galloway Bell, "The Next Four Years: What Lies Ahead?", <u>Construction Business Review</u>, (Vol. 3, No. 1), January/February, 1993.

Mahoney, William D., (Editor-in-Chief), <u>Building News Mechanical/Electrical 1993 Costbook</u>, BNI Building News, Los Angeles, California, 1993.

------, <u>Means Electrical Estimating - Standards and Procedures</u>, R.S. Means Company, Kingston, Massachusetts, 1986.

------, <u>Means Mechanical Estimating</u>, R.S. Means Company, Kingston, Massachusetts, 1987.

------, <u>Means Unit Price Estimating - A Comprehensive Guide</u>, R.S. Means Company, Kingston, Massachusetts, 1986.

McGeehin, Patrick A., "Identifying and Allocating Equipment Costs to Construction Projects," <u>Construction Business Review</u>, (Vol. 2 No. 6), November/December, 1992.

<u>Means Heavy Construction Cost Data 1993</u>, R.S. Means Company, Kingston, Massachusetts, 1993.

Merritt, Frederick S., (ed.), <u>Building Construction Handbook</u>, 3rd ed., McGraw-Hill, New York, 1975.

Mohorovic, Tiziana, (ed.), Business Failure Record - 1987, The Dunn & Bradstreet Corporation, New York, 1987.

------, Business Failure Record - 1988, The Dunn & Bradstreet Corporation, New York, 1988.

------, Business Failure Record - 1989, The Dunn & Bradstreet Corporation, New York, 1989.

Nunnally, S.W., <u>Construction Methods and Management</u>, 2nd ed., Prentice Hall, Englewood Cliffs, New Jersey, 1987.

Oxley, R. and P. Poskitt, <u>Management Techniques Applied to the Construction Industry - Fourth Edition</u>, BSP Professional Books, Oxford, England, 1986.

Paek, James H., "Common Mistakes in Construction Cost Estimation and Their Lessons", <u>Cost Engineering</u>, (Vol. 35, No. 6), June, 1993, pp. 29-33.

Page, John S., <u>Estimator's General Construction Man-hour Manual</u>, 2nd ed., Gulf Publishing Company, Houston, Texas, 1980.

------, <u>Estimator's Man Hour Manual on Heating, Air Conditioning, Ventilating and Plumbing</u>, 2nd ed., Gulf Publishing Company, Houston, Texas 1978.

------, and Jim E. Nation, <u>Estimator's Piping Man-hour Manual</u>, 3rd ed., Houston, Texas, 1979.

Palmer, William J., and William E. Coombs, <u>Construction Accounting and Financial Management - Fourth Edition</u>, McGraw-Hill, New York, 1989.

Peurifoy, R.L., <u>Estimating Construction Costs</u>, 4th ed., McGraw-Hill, New York, 1975

<u>Quality in the Constructed Project - A Guide for Owners, Designers and Constructors</u>, (Vol. 1), American Society of Civil Engineers, 1990.

Richardson General Construction Estimating Standards, (Volumes 1-3), Richardson Engineering Services, Inc., Mesa, Arizona, 1993.

Siddens, R. Scott: The Building Estimator's Reference Book, 23rd ed., Frank R. Walker Company, Lisle, Illinois, 1989.

Stewart, Rudney, Cost Estimating, John Wiley & Sons, New York, 1982.

Troy, Leo, Almanac of Business and Industrial Financial Ratios - 1993 Edition, Prentice-Hall, Englewood, New Jersey, 1993.

Tumblin, C.R., Construction Cost Estimates, New York, John Wiley & Sons, 1980.

Waeir, Phillip R., (Chief Editor), Means Cost Data 1993, R. S. Means Company, Kingston, Massachusetts, 1993.

------, Means Electrical Cost Data 1993, R. S. Means Company, Kingston, Massachusetts, 1993.

Walker, Anthony, Project Management in Construction - 2nd Edition, BSP Professional Books, Oxford, England, 1989.

Wetherill, Edward B., Estimating and Analysis for Commercial Renovation, R.S. Means Company, Kingston, Massachusetts, 1985.

Winslow, Taylor F., Construction Industry Production Manual, Craftsman Book Company, Carlsbad, California, 1972.

INDEX

Addenda 71
Allowances 24
Alternates 23
Ambiguity 169

Base Wages 37
Bid Bucket 72
Bid Bonds 15
Bid Day Preparation 70
Breach Of Contract 163
Buy Out 104

Change Orders 168
Changes 170
    Cardinal 170
    Constructive 170
    De facto 170
    Excessive 171
Claims 171
    Discrete Cost Approach 173
    Total Cost Approach 172
Close-Out 159
    As-Built Drawings 161
    Guarantees And Warranties 160
    Operations And Maintenance Manuals 161
    Punchlist 160
    Substantial Completion 160
Computers In Construction 191
Construction Industry
    Appeal 6
    Risks 4
Contract 163
Contractor Failures 1

Contracting Opportunities 13
Coordination Meeting 97
Cost Codes 132
Cost Control 124

Damages 165
Delays 174
    Compensable 174
    Concurrent 174
    Excusable 175
    Sequential 174
    Unexcused 174
Documentation 146
    Daily Field Report 149
    Progress Photos 147
    Tape Recorder 148
    Video Camera 149

Employee Benefits 37
Equipment Pricing 49
Estimate Spreadsheet 73
Estimating Errors 188
    Bid Day 189
    Judgment 189
    Pricing 189
    Quantity Survey 188
    Receiving Quotations 189
Estimator 7
Estimator's Database 34
Ethics 193
    Advance Payments 195
    Backcharges 195
    Bid Rigging 195
    Bid Shopping 194
    Idea Shopping 194
    Overbilling 195

    Substitutions 195

FICA 37

General Conditions Costs 59
    Clerical Help 60
    Employee Moving Expense 60
    Employee Subsistence 60
    Fuel 63
    General Purpose Labor 64
    Home Office Travel 60
    Job Signs 64
    Job Site Clean-Up 62
    Job Site Office Equipment And Furniture 64
    Job Site Safety 62
    Office Supplies 61
    Permits And Fees 65
    Personnel And Material Hoists 63
    Petty Cash 64
    Plans And Specifications 62
    Project Documentation 65
    Project Engineering 60
    Project Security 60
    Project Supervision 59
    Project Vehicles 63
    Protection Of Existing Conditions 63
    Protection Of New Construction 63
    Survey And Layout 62
    Temporary Heat 61
    Temporary Lighting 61
    Temporary Power 61
    Temporary Roads And Parking 65
    Temporary Toilets 61
    Third-Party Rentals 65

General Conditions Of The Contract 22

Instructions To Bidders 21, 27
Insurance 65
    Automotive Liability 68
    Builder's Risk 68
    Independent Contractor's Liability 68
    Owner's Protective Liability 68
    Product Liability 67
    Public Liability And Property Damage 66
    Unemployment 38, 39
    Workers' Compensation 38
Labor Pricing 43
Labor Productivity Factors 46-48
Labor Rates 37

Managing Risk 199
Material Costs 48
Mechanic's Liens 177
    Entitlement 178
    Filing 180
    Requirements 179
    Waiver 181
Meetings 145

Notices 177

Order Of Precedence 166
Owner - Contractor Agreement 22

Post Mortem To Bid 95
Prebid Conference 30
Profit Considerations 88
Progress Billings 150
Project Files 100
Project Management Errors 190
    Buy Out 190
    Claims 189

Notices	190
Progress Payments	191
Schedule	190
Subcontractors	191
Project Management Team	96
Project Manager	9
Project Superintendent	11
Proposal Form	21
Punchlist	93
Purchase Orders	112
Quality Assurance And Control	158
Quantity Survey	31
Requests For Information/Clarifications	119
Risk	200
Business	202
Insurable	203
Safety	121
Scheduling	
Initial	55
Project	101
Site Investigation	25
Special Conditions	23
Subcontractor/Suppliers	
Acoustical Ceiling	85
Asphalt Paving	81
Ceramic Tile	85
Concrete	82
Door Hardware	85
Drywall	86
Electrical	88
Elevator	86
Finish Carpentry	83
Fire Protection	87

    Glass 85
    HVAC 87
    Insulation 84
    Landscaping 81
    Masonry 82
    Nonperformance of 111
    Painting 86
    Payments to 110
    Plumbing/Piping 86
    Receiving Bids From 77
    Risk Factor 94
    Roofing 84
    Rough Carpentry 83
    Selection Of 108
    Site Concrete 82
    Site Grading 80
    Site Utilities 82
    Soft Flooring 86
    Soliciting Bids From 20
    Stucco 85
    Structural Steel 83
    Swimming Pool 86
    Waterproofing 84
    Wood Doors 84
Submittals 115
Summary Of The Work 23
Supplementary General Conditions 23
Surety 15
Surety Bond Cost 17

Taxes 90
Technical Specifications 24
Termination 164
Testing Costs 69
Training 196
Unit Prices 24, 53